Havasu Creek Dam site, September 14, 1923.

E. C. La Rue photograph, P.T. Reilly Collection, NAU.PH.97.46 Over size, USGS, courtesy of the Cline Library, Northern Arizona University.

DAMMING
GRAND CANYON

SCIENTIFIC AMERICAN

OCTOBER 1925

ALL IN THE DAY'S WORK

35¢ a Copy $4.00 a Year

Cover of *Scientific American* magazine, October 1925, showing an artistic fantasy of work during the 1923 USGS expedition in Grand Canyon. The artist was Howard V. Brown. The model for the topographer shown dangling from a cableway may have been Roland Burchard.

Courtesy of Scientific American.

DAMMING GRAND CANYON

The 1923 USGS Colorado River Expedition

Diane E. Boyer and Robert H. Webb
U.S. Geological Survey

Foreword by
Michael Collier

Utah State University Press
Logan, Utah

Copyright © 2007 Utah State University Press
All rights reserved

Utah State University Press
Logan, Utah 84322-7200

Manufactured in the United States of America
Printed on recycled, acid-free paper

ISBN: 978-0-87421-660-8 (cloth)
ISBN: 978-0-87421-665-3 (e-book)

Library of Congress Cataloging-in-Publication Data

Boyer, Diane E.
  Damming Grand Canyon : the 1923 USGS Colorado River expedition / Diane E. Boyer and Robert H. Webb.
     p. cm.
  Includes index.
  ISBN 978-0-87421-660-8 (cloth : alk. paper)
  1. Grand Canyon (Ariz.)–Description and travel. 2. Colorado River (Colo.-Mexico)–Description and travel. 3. Geological Survey (U.S.)–History–20th century. 4. Geological Survey (U.S.)–Biography. 5. River surveys–Colorado River (Colo.-Mexico)–History–20th century. 6. Dams–Colorado River (Colo.-Mexico)–History–20th century. 7. Water-supply–Political aspects–Arizona–Grand Canyon Region–History–20th century. 8. Water resources development–Arizona–Grand Canyon Region–History–20th century. 9. Grand Canyon (Ariz.)–Environmental conditions. 10. Colorado River (Colo.-Mexico)–Environmental conditions. I. Webb, Robert H. II. Geological Survey (U.S.) III. Title.
  F788.B69 2007
  917.91'320434–dc22
                                            2006103108

This book is dedicated to the past and present scientists who work in Grand Canyon, often envied but seldom understood; to those river runners who continue to stand vigilant for the canyon's resources; and to Steve Hayden and Toni Yocum.

Contents

Illustrations   *viii*

Foreword
  by Michael Collier   *x*

Introduction   *1*

1. Water and the Colorado Desert   *11*

2. Where Should the Dams Be? Politics, the Colorado River Compact, and the Geological Survey's Role   *32*

3. Prelude to an Expedition: Washington and Flagstaff   *48*

4. A Cumbersome Journey: Flagstaff to Lee's Ferry to the Little Colorado River   *90*

5. Surveys and Portages: Furnace Flats through the Inner Gorge   *121*

6. Of Flips and Floods: Bass Canyon to Diamond Creek   *164*

7. Feeling Their Oats: Diamond Creek to Needles   *209*

8. Aftermath: Politics and the Strident Hydraulic Engineer   *242*

About the Authors   *279*

Index   *280*

Illustrations

Front Endsheet   Havasu Creek Dam site
Back Endsheet   Specter Chasm Dam site
ii   Cover of *Scientific American* Magazine, October 1925
12   Map of the Colorado River Drainage
20   Map of the Lower Colorado River and the Salton Sink
21   The Western Side of the Salton Sink in 1905
35   Frederick Haynes Newell, First Director of the U.S. Reclamation Service
36   Arthur Powell Davis, Second Director of the U.S. Reclamation Service
59   Expedition Leader and Topographic Engineer Claude H. Birdseye
61   Head Boatman Emery C. Kolb
64   Boatman and Author Lewis R. Freeman
66   Boatman Leigh B. Lint
68   Boatman H. Elwyn Blake
69   Engineer Herman Stabler
70   Hydraulic Engineer Eugene C. La Rue
72   Geologist Raymond C. Moore
74   Topographic Engineer Roland W. Burchard
76   Rodman Frank B. Dodge
77   Cook Frank E. Word
77   Cook Felix Kominsky
79   Engineering Plans of the *Glen* from Drawings by Todd Bloch
88   Map of the Colorado River through Grand Canyon
91   Upstream View at the Mouth of Nankoweap Creek
92   Repairing the Road in Tanner's Wash, En Route to Lee's Ferry
95   The Crew at Lee's Ferry
95   Lee's Ferry, with the Vermillion Cliffs in the Distance
96   Soaking the *Marble* in the River at Lee's Ferry
96   Sketch of the Head of Marble Canyon and the Vermilion Cliffs near Lee's Ferry
108   Sketch of Marble Canyon from above Rapid 17
110   The Wreck of the *Mojave* at Cave Springs Rapid
111   La Rue with his Cameras at Vasey's Paradise in Marble Canyon
114   Surveying at 36 Mile Rapid
122   View Upstream of Granite Rapid from below the Mouth of Monument Creek

Illustrations

- 125 The Crew, Boats, and Camp Kitchen at a Camp at 75 Mile Canyon above Nevills Rapid
- 132 Edith Kolb and Leigh Lint in the *Boulder* at Hance Rapid
- 134 Repair of the *Boulder* at Hance Rapid
- 137 Claude Birdseye Records while Roland Burchard Surveys in the Upper Granite Gorge
- 139 Leigh Lint, Emery Kolb, and Elwyn Blake
- 147 Part of the Crew and Guests at the USGS Building at Bright Angel Creek
- 152 Downstream View of Hermit Rapid
- 156 Emery Kolb Rows the *Marble* through Hermit Creek Rapid
- 165 Eugene C. La Rue Measures Flow in Deer Creek below Deer Creek Falls
- 168 Upstream View of 128 Mile Rapid
- 189 Downstream View of the *Marble*, Wenched Up the Bank at Lava Falls Rapid
- 189 The *Marble*, Stranded High Above Water at Lava Falls Rapid
- 197 Sketch Map of an Unnamed Canyon on the Hurricane Fault Zone
- 203 Sketch Map of Diamond Peak in Western Grand Canyon
- 210 Travertine Falls at Mile 229 in Western Grand Canyon
- 218 Sketch of Separation Rapid
- 222 Emery Kolb Piloting the *Marble* through Separation Rapid
- 223 Sketch of Lewis Freeman and Frank Dodge
- 224 Sketch of Lava Cliff Rapid
- 227 Sketch of Leigh Lint
- 228 The Portage of Lava Cliff Rapid
- 230 Herman Stabler and Claude Birdseye with the Radio Setup at Devils Slide Rapid
- 232 The Ruins of Pearce Ferry, Now Submerged beneath Lake Mead
- 236 Boulder Canyon Dam site
- 238 Packing the Boats and Equipment at Needles, California
- 243 La Rue, Birdseye, and Stabler Standing near the *Grand*
- 256 Blake, Kominsky, and Lint at Diamond Creek
- 262 Sketch of Eugene C. La Rue
- 269 George Otis Smith, Fourth Director of the U.S. Geological Survey
- 275 Proposed Dam Sites along the Colorado River

Foreword

As a graduate student thirty years ago, I chose for my thesis a stretch of buckled rock along the Colorado River within Grand Canyon that could only be reached by boat. I carried the most up-to-date equipment—kapok-filled Mae West life preservers, a pocket calculator that could actually determine square roots, and plan-and-profile maps of the river corridor that Claude H. Birdseye had prepared fifty years earlier. I was a little troubled that the Birdseye maps were sprinkled with references to twenty-nine dam sites between Lee's Ferry and Black Canyon. But I wasn't too worried; hadn't David Brower and his Sierra Club saved the Grand Canyon from the dam-builders' antics in the 1960s?

How quickly we forget. At the turn of the twentieth century, entrepreneurs in the American Southwest were breathless with the possibilities of harnessing nature and harvesting its bounty. Surely the deserts would bloom if only crops could be sprinkled with holy water from the untapped Colorado River. Itinerant schemers and voluble visionaries lined the banks of the river, waiting for their chance to send water to California's Imperial Valley, to vegetable fields in Mexico, or even all the way to Los Angeles. But the Colorado had proven unruly with its track record of untamed floods punctuated by intermittent drought. Early attempts at water diversion were scuttled by flimsy headgates and fishy finances, stymied by disastrous flooding of the very lands that developers had billed as the future of a new West.

What was needed was a plan to control the river. And a plan required knowledge of the topography through which the river flowed. Maps in the middle of the nineteenth century had remarkably little to say about the canyons of the Colorado River. In 1869 and 1871–72, John Wesley Powell had successfully navigated much of the Green and Colorado rivers from Wyoming to Arizona and California, but his accounts were peppered with far too much hyperbole to be useful to engineers. By 1890, Robert Brewster Stanton had surveyed a hypothetical railroad through Cataract, Glen, and Grand canyons. The railroad would never be built, but the survey alone offers insight into the busy-beaver fervor of these early explorers: they truly believed that anything was possible.

Technological achievements never happen in a cultural or political vacuum. Railroads that had so recently stitched together the Atlantic and Pacific coasts were stoked as much by coal as by the religion of Manifest Destiny. The explosion of agriculture across the continent had been ignited as much by a Jeffersonian belief in the dignity of farming as by a fly-blown optimism that rain always follows the plow. Powell had tried, courageously and unsuccessfully, to link agricultural development of the West with the availability of water. But he was shouted down by self-interested speculators who were selling empty dreams, not sustainable communities.

The United States Reclamation Service, drawn like a rib from the U.S. Geological Survey in 1907, was responsible for water development throughout the nation. The USGS retained its keen interest in the science of water and rivers, especially in the arid Southwest. By the early 1920s, the Reclamation Service had focused on a few specific dam sites for the lower Colorado River, while the Geological Survey continued to peripatetically collect data about stream flow and canyon morphologies throughout the West. Both agencies were all too aware of the urgencies being created by the explosive growth of Southern California, with its appetite for water and power. In 1922, sharp knives drawn from all seven states bordering the Green and Colorado rivers had carved the region into upper and lower basins, unknowingly overallocating its resources at the outset of distribution.

Private power and water interests had been surveying the Colorado River system since the turn of the century. While the Reclamation Service somewhat prematurely insisted on a one-dam solution for control of the river, the Geological Survey thought instead to survey wider reaches of the river system—the San Juan River in 1921 and Green River, Glen Canyon, and the lower Colorado in 1922. The crowning jewel in this series of explorations would be the Geological Survey's 1923 expedition through Grand Canyon, led by Claude Birdseye.

Birdseye, an athletic 45-year-old distant cousin to the world of pot pies, turned out to have all the requisite personality traits that would be required to hold this expedition together. His crew consisted, among others, of a destructively headstrong hydrologist, a capable but often petty head boatman, and one true hero who happened to be an alcoholic. Two months and nineteen days after beginning their voyage at Lee's Ferry, Birdseye and his men would emerge from Grand Canyon not only having met the not-inconsiderable demands of life on the water but having also accurately mapped the Colorado River and relevant stretches of its bedrock geology. Beyond being survivors, they were scientists. Despite rotten-bottomed boats, a conflicted crew, and a temperamental river, the 1923 expedition was able to accurately observe the Grand Canyon and commit those observations to a map whose integrity I could still rely upon fifty years hence.

Who were these people who dared dream of damming the Colorado? They were geologists, hydrologists, and engineers who were products of their day. Herman Stabler was a topographer who joined the trip midway, near Phantom Ranch. Upon reaching the river, Stabler wasted few words setting the scene before plunging into the first of many dam site descriptions in his diary. Four days later, he loosened up a little, commenting that he "enjoyed riding the rapids." The following day, Claude Birdseye sighed for just a second and then acknowledged that his proposed dam site at Ruby Rapids might have engineering drawbacks, but at least it wouldn't inundate Bright Angel Creek and Phantom Ranch. Beyond this brief reflection, no members of the party expressed concerns about the havoc that their imagined dams would wreak. I shudder to think of a Grand Canyon reduced to a series of tubs and spigots, but is it reasonable to impose twenty-first-century environmental judgments upon these early-twentieth-century men? They were firmly rooted in the American values and enthusiasms of their time. Let them dream of their dams.

They were scientists, though, if not immediately recognizable to everyone today. After all, didn't the crew just tinker with alidades, scramble over cliffs, and scribble lines on soggy paper? A better question would be how has geologic science changed over time at Grand Canyon? John S. Newberry arrived first on the scene in 1858, and appreciated that the canyon was a world-class showcase for the powers of erosion. What a scientist Newberry was, and, man, could he grow a beard! Powell followed a decade later, hypothesizing about down-cutting of the river and uplift of the Colorado Plateau. His right-hand-man, Clarence Dutton, understood that nearby higher formations had once covered the region before being stripped off. Dutton published these thoughts in 1882 in that most eloquent of scientific treatises, *The Tertiary History of the Grand Cañon District*. Geology to these pioneers was a descriptive sport. The science would begin to branch out in the twentieth century, though.

Eliot Blackwelder and Chester Longwell wrestled with questions about evolution of the Colorado River's course through Grand Canyon. Eddie McKee systematically measured, defined, and published interpretations of almost all of the canyon's sedimentary layers during the nineteen forties, fifties, and sixties. George Billingsley first huffed and puffed through the canyon while trying to keep up with Harvey Butchart in 1966; since then he, more than anyone else, has single-handedly mapped all of Grand Canyon's geologic strata and structure. Since the 1980s, flocks of young academics, like Karl Karlstrom, have perched on ledges throughout the canyon, pecking at its Precambrian basement while trying to fit the Vishnu Schist and Zoroaster Granite into a larger tectonic framework. Ivo Luchitta investigated adjacent sedimentary basins and overlying basalts, trying to understand how and when the river melted through these rocks. Bob Webb focused geologic attention on the role of debris flows that can obstruct the river and twist it into rapids.

How does the Birdseye expedition fit into this parade of science? Snugly, I would say. Early geologists were explorers. The next wave—including Birdseye—were mappers. Then came interpreters, followed by theoreticians. One last approach to canyon science remains in this parade, but we will get to that in a moment. The Birdseye expedition was charged with answering a single question: where *can* dams be built? The obvious corollary question—where *should* dams be built?—existed in a parallel universe, to be answered, not by Birdseye, but by society at large. One member of the expedition, hydrologist Eugene La Rue, made the unpardonable mistake of taking the answer upon himself.

La Rue believed that, in addition to a handful of smaller structures, high dams at Bridge Canyon and at the foot of Glen Canyon would best serve the water needs of the Southwest. He had not counted on the power of already-vested interests in Los Angeles and Washington D.C. to hold sway over his impeccable scientific logic. A decade after the 1923 expedition, La Rue watched in dismay as Hoover Dam was constructed in a location he did not favor. La Rue died in 1947, nine years before construction would begin at Glen Canyon Dam, a massive concrete arch whose design incorporated some of his original ideas. In defeat, La Rue had failed to understand that, in the rigged race between science and public policy, it's safer to put your money on politicians than scientists.

A "new" type of science has turned up at Grand Canyon in the last few years, one that feeds on policy and is funded by water and power interests. Glen Canyon Dam is pretty good at holding back water, and very good at generating spikes of electrical power. Since 1982, biologists, archaeologists, and geologists have descended upon the shores of the Colorado like hordes of grasshoppers, pulling up every hapless tamarisk they can find, and analyzing every sand grain that hasn't yet been flushed downstream by the dam's variable releases of clear water. These men and women are committed to Grand Canyon, convinced that they have undertaken an important mission, and confident that they can make a difference. And yet, a quarter century and hundreds of millions of dollars later, they haven't been allowed to save a single endangered species or forestall the deflation of a single beach. Politics and power win again. Perhaps theirs is not really a new science, though. After all, none of the dam sites surveyed by Birdseye was ever built either.

In the end, I am drawn to the individuals who made up the Birdseye expedition, rather than anything they did or did not accomplish. Raymond Moore, a true geologist, was ecstatic to discover ancient "calcareous algae" mats above Bass Camp. Felix Kominsky was the prototypic happy cook about camp, always ready to laugh or help. Birdseye, though focused on getting downriver, was always willing to declare a day "Sunday" if he sensed that his men were overtaxed. Best of all, I admire Frank Dodge who consistently put the expedition's success before any personal consideration. A powerful swimmer, he dove into the river to rescue head boatman Emery Kolb who had flipped in Upset Rapids. Dodge, like people with whom I would later commercially row, was a drifter, lost in his cups, and died before his time. But he was one of those bigger-than-life men, an appreciative realist who wrote, "Incidentally, Birdseye was the finest boss I ever had." Dodge understood the beautiful jealousy that all river runners feel for the Grand Canyon once we have seen her from the inside out.

Michael Collier
December 2006/Flagstaff

Introduction

Our interest in the 1923 expedition in Grand Canyon came from several sources. First and foremost, this U.S. Geological Survey (USGS) trip was the first that our agency sponsored in Grand Canyon. This is a notably poignant reason given that some USGS scientists, including the authors, have spent much of their careers in the challenging environment of the canyon, toiling to collect scientific data. Despite its significance, the 1923 trip is underrepresented in our agency's written history.[1] Second, USGS has a rather onerous review process of its publications, passionately attempts to stay nonpartisan, and avoids advocacy by its employees. A rather dim institutional memory holds that this policy was initiated in response to the aftermath of the 1923 expedition.[2] Finally, the 1923 trip, one of the great surveying exploits of the twentieth century,[3] yielded scientific data that form a cornerstone of our scientific research on

---

1. Robert Follansbee, "A History of the Water Resources Branch of the United States Geological Survey," vol. 2, "Years of Increasing Cooperation July 1, 1919 to June 30, 1928" (Reston, VA: U.S. Geological Survey, unpublished manuscript, n.d.), barely mentions the 1923 Grand Canyon expedition. The official history of the U.S. Geological Survey, by Mary C. Rabbitt, *Minerals, Lands, and Geology for the Common Defence and General Welfare*, vol. 3, *1904–1939* (Washington, DC: U.S. Government Printing Office, 1986), 243–44, devotes only one paragraph to this trip, and most of this discusses the death of President Warren Harding, a largely irrelevant incident during the expedition.
2. This institutional memory comes from the two published forms of Walter Langbein's article on La Rue and his effect on USGS publication policies. One form is in an obscure USGS publication, Walter B. Langbein, "L'Affaire LaRue," *U.S. Geological Survey Water Resources Division Bulletin* (April—June 1975): 6–14. The other is in a historical journal, Walter B. Langbein, "L'Affaire LaRue," *Journal of the West* 22 (1983): 39–47. The articles are similar but are not identical. We quote from both of them, and use the date of publication to differentiate the two versions.
3. According to a motion picture made about the expedition many years later, it was "the most difficult survey ever undertaken by government engineers." Fred Watkins (producer), *The 1923 Surveying Expedition of the Colorado River in Arizona by the United States*

long-term ecological and geomorphic change in Grand Canyon.[4] Even though the leaders of this trip published their accounts,[5] and others have written of it as well,[6] we felt that the story needed to be retold from the perspective of several participants whose voices have not been heard but whose lives were profoundly affected.

The 1923 expedition cannot be considered without placing it within the water-development framework of the American West. Beginning with John Wesley Powell, the second director of USGS, science and politics have clashed over the Colorado River and the development of its water resources. Powell's controversial book, *Report on the Lands of the Arid Region of the United States*,[7] advocated

---

Geological Survey (Washington, DC: The American Society of Photogrammetry and Remote Sensing, produced by the Northern California Region, 1977)

4. Raymond M. Turner and Martin M. Karpiscak, *Recent Vegetation Changes Along the Colorado River Between Glen Canyon Dam and Lake Mead, Arizona* (U.S. Geological Survey Professional Paper 1132, 1980); Theodore S. Melis, Robert H. Webb, Peter G. Griffiths, and Tom J. Wise, *Magnitude and Frequency Data for Historic Debris Flows in Grand Canyon National Park and Vicinity, Arizona* (U.S. Geological Survey Water Resources Investigations Report 94–4214, 1994); Janice E. Bowers, Robert H. Webb, and Renee J. Rondeau, "Longevity, recruitment, and mortality of desert plants in Grand Canyon, Arizona, U.S.A.," *Journal of Vegetation Science* 6 (1995): 551–64; Christopher S. Magirl, Robert H. Webb, and Peter G. Griffiths, "Changes in the water surface profile of the Colorado River in Grand Canyon, Arizona, between 1923 and 2000," *Water Resources Research* 41 Wo5021, doi:10.1029/2003WR002519 (2005).

5. Claude H. Birdseye, "Surveying the Colorado Grand Canyon," *The Military Engineer* 16 (1924): 20–28; Claude H. Birdseye and Raymond C. Moore, "A boat voyage through the Grand Canyon of the Colorado," *Geographical Review* 14 (1924): 177–96; Lewis R. Freeman, *Down the Grand Canyon* (London: William Heinemann, 1924); Lewis R. Freeman, "Surveying the Grand Canyon of the Colorado," *The National Geographic Magazine* 45 (1924): 471–530, 547–48. Birdseye also was involved in the preparation of an oft-used press release and gave a radio address: "Surveying the Grand Canyon, Ruggedest 300 Miles of the Canyon of the Colorado Traversed by Geological Survey Party" (Washington, DC: U.S. Geological Survey Library, November 11, 1923); Claude H. Birdseye, "Radio in the Grand Canyon" (Washington, DC: Station WRC of the Radio Corporation of America, March 21, 1924), Accession 71541, Smithsonian Institution, Washington, DC.

6. Brief coverage of the expedition is found in David Lavender, *River Runners of the Grand Canyon* (Grand Canyon, AZ: Grand Canyon Natural History Association, 1985), 58–65; Richard E. Westwood, *Rough-Water Man: Elwyn Blake's Colorado River Expeditions* (Reno: University of Nevada Press, 1992), 124–223; Donald L. Baars and R. C. Buchanan, *The Canyon Revisited: A Rephotography of the Grand Canyon 1923/1991* (Salt Lake City: University of Utah Press, 1994). In his comprehensive treatment, P. T. Reilly discusses the trip in the context of the overall history of Lee's Ferry. P. T. Reilly, *Lee's Ferry: From Mormon Crossing to National Park*, ed. R. H. Webb (Logan: Utah State University Press, 1999), 275–302. William Suran also includes a chapter in his online biography of brothers Emery and Ellsworth Kolb; William Suran, "With the Wings of an Angel: A Biography of Ellsworth and Emery Kolb, Photographers of Grand Canyon," http://www.kaibab.org/kolb/index.html (accessed September 29, 2004).

7. John Wesley Powell, *Report on the Lands of the Arid Region of the United States*, ed. Wallace Stegner (Washington, DC: U.S. Government Printing Office, 1879; Cambridge,

regional development centered upon drainage basins and a realistic evaluation of water resources; the resulting firestorm resulted in Powell's resignation from USGS in 1894.[8] USGS subsequently chose an institutional course towards water-resource evaluation without resource development or management responsibility, leaving the latter to the U.S. Reclamation Service (later the U.S. Bureau of Reclamation), a spinoff agency. The stakes were high, fueled by interagency and interpersonal competition. There were very public clashes between the Director of the Reclamation Service, Arthur Powell Davis, and Eugene C. La Rue, the hydraulic engineer on the 1923 expedition, which culminated in contentious testimony before the U.S. Senate in 1926. One result, following extended public debate, was construction of Hoover Dam, one of the great engineering accomplishments of the twentieth century. Another was the relegation of the Water Resources Division of the U.S. Geological Survey to a reduced role of hydrologic data collection, a role that persisted for many decades.[9]

Sources of the Diaries and Other Materials

We used the diaries and letters written by the participants to form the backbone of this book. Of the twelve men involved, eight—Claude Birdseye, Elwyn Blake, Emery Kolb, Lewis Freeman, Eugene La Rue, Leigh Lint, Raymond Moore, and Herman Stabler—wrote journals for the entire or part of the trip that have survived the intervening years. Roland Burchard did not keep a diary but wrote letters; Frank Dodge wrote an abbreviated account in his autobiography; and if Frank Word and Felix Kominsky kept diaries, they are not publicly available. We collated a complete set of these diaries and other narrative material, and for the sake of brevity in this book, we edited the diary entries. In most cases, we corrected spelling; grammar is corrected where appropriate; and in other cases, we present information verbatim. We then silently reduced the entries to tell one consistent story from one or, where deviations occurred, multiple perspectives.

We obtained a copy of Birdseye's account from the National Archives and Records Administration (NARA) in College Park,

---

Massachusetts: Belknap Press, 1962). Powell's seminal report has been reprinted as John Wesley Powell, *The Arid Lands* (Lincoln: University of Nebraska Press, 2004).
8. Donald Worster, *A River Running West: The Life of John Wesley Powell* (New York: Oxford University Press, 2001).
9. Langbein, "L'Affaire LaRue" (1975):6.

Maryland, which holds both his original handwritten diary as well as a typed copy.[10] NARA also holds typescripts of the diaries of Freeman, Stabler, Blake, and Lint, as well as a photocopied set of Blake's serialized diary as it appeared in the *San Juan Record*, a Monticello, Utah, newspaper, under the title "Diary of a Voyager on the Colorado River."[11] When Ana MacKay first entered the Birdseye, Freeman, Stabler, Blake, and Lint diaries into a computer in the mid-1990s, she worked from the NARA typescripts. We compared Birdseye's handwritten original, obtained in 2006, with the typescript, which apparently is a verbatim transcription. Likewise, we compared the NARA typescript of Blake's diary against the original *San Juan Record* columns; they varied by only a few typographical errors.

The Huntington Library in San Marino, California, holds three versions of Blake's diary within the huge collection amassed by river historian Otis "Dock" Marston, all based on Blake's original handwritten journal, which was destroyed in a flood. The first is a photocopy of the NARA typescript. The second is a typescript prepared by Kathryn Gore in 1950 for Marston, working from Blake's faint, difficult-to-read pencil original, and is virtually identical to the NARA document. Blake may have typed the third version himself in 1947, working from his original. Marston noted: "In making the copy, Blake has made extensive revision in the arrangement of phrases and has added some items which serve to clarify his meaning. The editing does not change the meaning or character delineation." We chose to use Blake's published version as it was composed by Blake soon after the expedition, without the possible errors introduced by a third party attempting to read the original years later. Blake's unpublished autobiography is also housed in the Marston Collection; we include a few quotes from it and reference it in the endnotes.[12] Because much of Blake's story has already

---

10. Claude H. Birdseye, "Diary of Grand Canyon Survey 1923" (College Park, Maryland: National Archives and Records Administration, Record Group 57, Records of the Topographic Division). Most of the materials relating to the 1923 expedition are held in three boxes clearly identified in the finding guide for Record Group 57.
11. The *San Juan Record* prefaced each episode of Blake's diary, "In publishing these happenings we feel that the paper is giving its readers on opportunity to get an idea of the stupenduousness of the great canyon, and at the same time get an idea of the mammoth work which the government is carrying on with a view of ultimately harnessing this unruly and devastating stream."
12. All three versions of Blake's diary are in box 21, folder 1, Otis "Dock" Marston Collection, Huntington Library, San Marino, California. His autobiography is "As I Remember," unpublished manuscript, box 430, folder 1, Marston Collection.

been told, we emphasize other voices in our account except where Blake's experiences are front and center.[13]

In 2005, we discovered Freeman's original bound diary, handwritten[14] in pencil in a barely legible, small script, at the Huntington Library, within the Marston Collection. After the trip, Freeman (or possibly someone else, but that is unlikely given the challenge in deciphering the original) typed the edited version, a copy of which we had obtained from NARA. We chose to use this typescript as our primary source due to the similarity of the two documents, as a nod to Freeman as a published author, and to make it parallel with the bulk of the other diaries—which were also typed and presumably, edited, versions of the originals. The handwritten diary and typed manuscript are remarkably similar, although there are some significant differences in prose and detail of description, most of which do not change meaning. In a few cases, Freeman deleted large passages or verbose descriptions. When we felt these provided an important perspective that would otherwise be missing, we replaced his typescript with these entries and designate these excursions as "Freeman, diary."

La Rue was the official trip photographer, taking still and moving pictures, much to Kolb's disgust (as we discuss in Chapter 3). La Rue's original handwritten diary, with a photo of his wife and three daughters pasted inside the front cover, is housed in the La Rue manuscript collection at the Huntington Library in box 10, folder 1. A separate bound journal contains La Rue's detailed photographic notes. The La Rue collection includes numerous other papers associated with the 1923 expedition, La Rue's fall from USGS grace, and his post-government career. During the 1923 expedition, La Rue may have quit writing after his entry for Clear Creek, or the diary may have been lost in Separation Rapid as other diary entries imply (see Chapter 7). Rosalyn Jirge transcribed the extant La Rue diary, and we have largely worked from her copy, with corrections from the original. Mabel La Rue's accounts of her visit to Phantom Ranch are also in the La Rue papers at the Huntington Library. La Rue's last name has various spellings; many sources, including the original diaries, spell it LaRue or Larue, creating some confusion. Eugene spelled it La Rue, at least most of the

---

13. Westwood, *Rough-Water Man*.
14. Freeman hand wrote his diary; he did not, as previously claimed by some authors, type it on the first typewriter to be hauled through Grand Canyon. There was no typewriter on the 1923 expedition. Lewis Freeman, unpublished diary, box 69, Marston Collection.

time, and that is the spelling we use. His photographs are stored in the U.S. Geological Survey Photograph Library, but, as described below, we obtained his edited motion picture from other sources.

Our first copy of Lint's diary was a typescript from NARA, and we found an identical copy in the Kolb Collection at the Cline Library at Northern Arizona University (NAU) in Flagstaff, Arizona. Leigh is pronounced "Lee," and when diarists spelled his name that way, we corrected these entries. George Lint, Leigh's son, later loaned us his copy of the diary (also a typescript like those held at NARA and NAU), and we reproduced Lint's sketches from this version, which contains original pen-and-ink drawings. George Lint also provided us with photographs, correspondence, and other items relating to Lint's career, as well as a partial set of copies of Blake's serialized diary, and a copy of the edited USGS expedition film.[15] Another version of this film is in the Marston Collection at the Huntington Library but, when this project was started, was considered to be too fragile to be viewed. In 1977, the American Society of Photogrammetry and Remote Sensing in Washington, D.C produced an edited version of the 1923 expedition film, featuring a soundtrack and a brief commentary by Leigh Lint.[16]

Preston Burchard, Roland's son, provided his father's materials for our use; Preston's birth in 1923 delayed his father's arrival at the start of the expedition. Dodge wrote a brief account in his autobiography, *The Saga of Frank B. Dodge*,[17] published as a USGS administrative report but not widely distributed.

Kolb's original diary, which, like La Rue's, ends early—but at the Bright Angel Creek confluence—is housed at NAU, which also archives other papers and photographs produced by the Kolb Studio, in manuscript 197, box 17, folder 20. If Kolb kept a second diary below Bright Angel, its whereabouts are unknown or it was lost, perhaps in his flip in Upset Rapid. On its Web site,[18] NAU hosts the Kolbs' *Grand Canyon Film Show*, which includes some footage of

---

15. The 1923 USGS expedition film was converted to digital video, and the original is now stored at the Cline Library, Northern Arizona University, Flagstaff, Arizona. Copies of the film were deposited with the U.S. Geological Survey. *The Survey of the Grand Canyon of the Colorado River in Arizona*. U.S. Department of the Interior: United States Geological Survey, no date. A copy of Lint's diary typescript, photographs, and other papers pertaining to his river running career are also at the Cline Library.
16. Watkins, *The 1923 Surveying Expedition*.
17. Frank B. Dodge, *The Saga of Frank B. Dodge* (Tucson, AZ: U.S. Geological Survey Administrative Report, 1944).
18. Emery and Ellsworth Kolb, *Grand Canyon Film Show* (Grand Canyon, AZ: Kolb Studio, no date), http://www.nau.edu/library/rm/kolb (accessed February 15, 2006).

the 1923 expedition. Edith Kolb, Emery's sixteen-year-old daughter, kept a diary of her time with the men at Lee's Ferry prior to their departure, written in a flowery hand and illustrated with wonderful sketches by Moore. It is also in the Emery Kolb Collection, in box 14, folder 1756. We quote portions of it that illuminate the character of the expedition's participants. The original is with the Kolb Collection at NAU. William Suran's online biography "With the Wings of an Angel"[19] is by far the most detailed textual material regarding the 1923 trip, although it is distinctly biased toward Kolb's version of the expedition.

Moore's original journal, which consists primarily of geologic notes and sketches, is housed at the University of Kansas Kenneth Spencer Research Library.[20] This journal was featured in another account of the 1923 expedition[21] and is not emphasized here. Stabler's account begins with his arrival at Phantom Ranch in the middle of the expedition. Our only Frank Word document, a letter written to La Rue after the trip, is from the La Rue Collection at the Huntington Library.[22] We were unable to locate a single document written by Felix Kominsky.

Correspondence relating to the trip's preparation and its aftermath is abundant within the Kolb Collection at NAU, the Marston and La Rue Collections at the Huntington Library, and, to a lesser extent, in the USGS records at NARA. We obtained personnel files for Birdseye, Burchard, Dodge, La Rue, Lint, and Moore from the National Personnel Records Center in St. Louis, Missouri, and these documents provided many details of their careers that shed light on their personalities and achievements.

In editing the diaries and letters, our main goal was to tell the story in a complete but readable fashion, letting the diarists speak for themselves. We give nods to two other first-person accounts of historical river trips as our inspiration to collate diaries in chronological order.[23] With a few exceptions, we use only one entry to

---

19. Suran, "With the Wings of an Angel."
20. Raymond Moore, unpublished diary, box 10, Raymond C. Moore Collection, University Archives, PP 3, Kenneth Spencer Research Library, University of Kansas Libraries.
21. Baars and Buchanan, *The Canyon Revisited.*
22. Box 2, folder 1, La Rue Collection, Huntington Library, San Marino, California.
23. John Cooley, *The Great Unknown: The Journals of the Historic First Expedition Down the Colorado River* (Flagstaff, AZ: Northland Publishing, 1988) provided the inspiration for collating different voices in chronological order. We reserve our highest respect for D. L. Smith and C. Gregory Crampton, *The Colorado River Survey, Robert B. Stanton and the Denver, Colorado Canyon and Pacific Railroad* (Salt Lake City, UT: Howe Brother Books, 1987) for instilling in us the value of diaries and original voices.

recount the details of any given event. For some notable events, such as boat flips or floods, we used the voices of several diarists to paint a view from divergent angles. We attempted to include words composed by each participant. Our editorial notes are denoted in brackets or in footnotes. We have included selected excerpts from items of questionable accuracy, such as memoirs written many years after the fact and newspaper accounts, because they offer a different perspective.

Heavily sanitized versions of the expedition appeared in print within the two years following the expedition.[24] Birdseye, as the expedition leader, gave several lectures, as did other expedition members. Freeman, the official expedition chronicler, authored a number of articles and *Down the Grand Canyon*, a book that presents his account of the expedition, and which has been heavily drawn upon in more recent histories.[25] La Rue published his technical findings in 1925 and vociferously defended his water-development plan, as we report.[26] We briefly follow him through his later career as well.

We choose to depart from the official place names used by the Board on Geographic Names to denote geographic features we discuss. Since Lee's Ferry honors John D. Lee, a historic figure in southern Utah history, we choose to use the possessive in referring to this place. We correct other misspelled or otherwise incorrect names, such as Dubendorff Rapid (for Seymour Dubendorff but misspelled Deubendorf Rapid by the Board on Geographic Names). With the exception of Lee's Ferry, we provide the corrections in footnotes or in parentheses in the text.

Acknowledgments

We thank a large number of people who inspired, contributed, or otherwise helped with this book. Robert Hirsch, chief hydrologist of the U.S. Geological Survey (USGS), greatly encouraged this

---

24. Birdseye, "Surveying the Colorado Grand Canyon;" Birdseye and Moore, "A boat voyage through the Grand Canyon;" Freeman, "Surveying the Grand Canyon of the Colorado."
25. Westwood, *Rough-Water Man*, borrows greatly from Freeman's and Birdseye's writings in the chapter on Elwyn Blake's exploits in Grand Canyon.
26. Eugene C. La Rue, "Tentative plan for the construction of a 780-foot rock-fill dam, on the Colorado River, at Lees Ferry, Arizona," *Transactions of the American Society of Civil Engineers* 86, Paper 1512 (1923): 200–267; Eugene C. La Rue, *Water Power and Flood Control of Colorado River below Green River, Utah* (Washington, DC: U.S. Government Printing Office, U.S. Geological Survey Water-Supply Paper 556, 1925).

book in its earliest stages. He pointed us toward the work of Walter Langbein, who documented the effect that La Rue had on the policies of what was then the Water Resources Division of USGS. We especially thank the remarkable Ana MacKay, who laboriously typed the diaries, in some cases from barely legible manuscripts. Several USGS colleagues, including Colleen Allen of the Central Records Library and Carol Edwards of the Field Records Library (in Denver, Colorado), Elizabeth Colvard of the Earth Science Information Center in Menlo Park, California, and Clifford Nelson in Reston, Virginia, provided a great deal of useful information and suggestions. John Gray unearthed Dodge's autobiography from the dusty files of the Arizona District in Tucson.

We offer our considerable thanks to George Lint and the late Preston Burchard, who graciously loaned us materials belonging to their fathers, who were key and, until now, unsung participants on the 1923 expedition. George Lint, in particular, kept one of the few remaining copies of the edited film footage from the expedition. Without their contributions, our knowledge of the contributions of Roland Burchard and Leigh Lint would be greatly diminished. We thank Dan Cassidy of Five Quail Books for putting us in touch with Preston Burchard and for pointing us to other useful sources.

The dependably capable staff at the Northern Arizona University Cline Library Special Collections and Archives Department enthusiastically provided access to their collections and supported this project. We particularly acknowledge the assistance of Richard Quartaroli, Karen Underhill, Bee Valvo, and Susan McGlothlin. Paula Johnson of the Smithsonian Institution shed light on the *Grand*, the boat rowed by Lewis Freeman. We gratefully acknowledge the assistance of the staff of the Huntington Library, which houses the massive Otis "Dock" Marston and E. C. La Rue collections. The Marston Collection, in particular, is an invaluable source for any student of Grand Canyon history. Joe Schwartz of the National Archives and Records Administration helped us with the official USGS records held there. Rosalyn Jirge kindly shared her research notes obtained during her visits to the Huntington. Becky Schulte and Kathy Lafferty of the Kenneth Spencer Research Library, University of Kansas, helped us obtain a copy of Raymond Moore's diary. Likewise, whether he knows it or not, John Weisheit is always an inspiration. Richard Quartaroli provided some of the obscure details that enrich any book and corrected several factual errors. Archaeologist Gary Huckleberry shared his knowledge of prehistoric Hohokam canals in Arizona. Brad Dimock provided insights into Bert Loper and his role in USGS expeditions. Lew Steiger answered some of our questions about the Kolbs. Keith Howard provided stimulating discussion on Lake Cahuilla.

The online, searchable versions of the *New York Times*, *Washington Post*, and especially the *Los Angeles Times* were invaluable references. Surprising amounts of material were available on the Internet, much of it posted by archives and libraries. We relied heavily on the resources of the University of Arizona Library, particularly its inter-library loan service. The G. E. P. Smith papers, housed in Special Collections at the University of Arizona, provided additional insights into Eugene C. La Rue.

We thank Preston Burchard, Brad Dimock, Thomas Hanks, Robert Hirsch, George Lint, Christopher Magirl, Richard Quartaroli, Bennett Raley, W. R. Rusho, and John Weisheit for critically reviewing parts or all of our manuscript. Stephen Hayden kindly endured endless descriptions as the draft took place, making many valuable suggestions on how to improve the book.

Information on Note Citations

Items from the Marston and La Rue Collections are courtesy of the Huntington Library, San Marino, California; Kolb Collection materials are courtesy of the Special Collections and Archives Department, Cline Library, Northern Arizona University, Flagstaff, Arizona; all of the personnel files are courtesy of the National Personnel Records Center in St. Louis, Missouri (we include only the last names of the men in referencing them); NARA is used for the National Archives and Records Administration in College Park, Maryland. Citations from the *Los Angeles Times*, *Washington Post*, and *New York Times* were obtained via the ProQuest online searchable newspaper databases, accessed through the University of Arizona Library's subscription.

1

Water and the Colorado Desert

At the dawn of the twenty-first century, the Colorado River well deserves its nickname as "the American Nile," first suggested in the early years of the twentieth century.[1] One of the most regulated watercourses in the world, and certainly in the United States,[2] each drop of Colorado River water is reputedly reused five times before it reaches the ocean. It is used so thoroughly that only a small percentage still reaches the Sea of Cortés. Most of the runoff from the headwaters evaporates from large reservoirs, evaporates or transpires from agricultural fields, is diverted to the large metropolises in the West, flows into the Salton Sea as agricultural wastewater, infiltrates into regional groundwater, or is discharged as effluent into the Pacific Ocean off the California coast. Along the border between California and Arizona, the Colorado is more engineered canal than natural river; upstream, reservoirs large and small store, regulate, and re-regulate its precious waters.

Even with all the flow regulation, water delivery is not guaranteed. Compact Point, the figurative point of division between the upper and lower basin states that use the Colorado River, is one mile downstream from the mouth of the Paria River at Lee's Ferry.[3] Fifteen and one-half miles upstream from Lee's Ferry stands Glen

---

1. Daniel T. MacDougal, "The delta of the Rio Colorado," *Bulletin of the American Geographical Society* 38 (1906): 1–16.
2. R. M. Hirsch, J. F. Walker, J. C. Day, and R. Kallio, "The influence of man on hydrologic systems," in *The Geology of North America*, vol. O–1, *Surface Water Hydrology*, ed. M. G. Wolman and H. C. Riggs (Boulder, CO: Geological Society of America, 1990), 329–59.
3. In Colorado River Compact legalese, Lee's Ferry is specified as Lee Ferry.

The Colorado River drainage.

Canyon Dam. On January 1, 2005, the water level in Lake Powell, the reservoir impounded by the dam, stood at 3,564 feet above mean sea level. This statistic on one of the world's largest and most important reservoirs seems benign enough until it is accompanied by something more meaningful: the lake level was 135 feet below full-pool elevation, and the amount of water stored in the lake on that date was only 35% of capacity. Roughly six years of drought had decreased inflows into the reservoir, and the mandated discharge of 8.23 million acre feet of water every year into the lower basin had driven lake levels to unprecedented lows. On January 27, the reservoir held 8.51 million acre-feet of water, just a little more than one year of mandated flow releases. Other reservoirs in the Colorado River drainage were in a similar condition.

Water managers responsible for use of the Colorado River, from local water districts to the head of the Department of Interior, were highly concerned with not only the reservoir levels but also some of the long-term prognostications concerning the magnitude and length of the drought. The media were filled with reports and updates, and the managers began, for the first time, to discuss long-term water conservation measures. The scientific predictions as to the potential length of the drought were wide ranging, with the most dire holding that the drought could last another twenty to thirty years.[4] If the drought were to persist that long, by its end, Lake Powell would impound no water, serving instead as a minor impediment to delivery of water from snowy headwaters to thirsty lower basin. Some hope came in the form of a developing wet winter; as of January 27, 2005, overall snowpack conditions within the Colorado River drainage were 125% of normal, suggesting the likelihood of above-average inflows into Lake Powell in the coming spring runoff. By the end of 2005, water levels in Lake Powell were at 50% capacity, but a return to severe drought in 2006 created more dire warnings of water shortages.

How users of Colorado River water got to this point is a story of political intrigue, the need for water to spur development, engineering know-how, and intrepid exploration. In the early 1900s, the problem was too much water at the wrong time. Farmers and politicians demanded a federal solution to the out-of-control, and hence unusable, river. Promoters of desert farms and future urban areas lusted for predictable, controlled water deliveries. Climatic fluctuations—both high runoff periods and drought—played a large role in spurring on water development with an overly optimistic estimate of water yield. Two federal government agencies, one an outgrowth of the other, jousted in a highly public forum for approval of a water plan for development of what arguably is the single most important water supply in the United States. Ultimately, politics prevailed over planning and science, literally and figuratively casting the Colorado River and delivery of its water resources in cement.

Our story centers on how potential dam sites were identified and how the dam site in Black Canyon was chosen over many others in the 1920s. We consider the role of the U.S. Geological Survey

---

4. K. M. Schmidt and R. H. Webb, "Researchers consider U.S. Southwest's response to warmer, drier conditions," *EOS* 82 (2001): 475, 478.

and a largely forgotten expedition through Grand Canyon, one led by a wizened surveyor and manned by a motley crew of scientists and mostly novice boaters. A head boatman who thought he owned the rights to all photographic images of Grand Canyon guided the expedition. One of the boatmen was an overweight adventure writer, two others were young boaters, and perhaps the most valuable of all was the rodman whose boat sank early in the expedition. The hydraulic engineer was better respected for his science than his temperament, the geologist eventually showed his world-class talents, and the topographers were steady career employees of the federal government. A nation fascinated with their courage and daring followed their exploits.

Before we get to the U.S. Geological Survey and its 1923 expedition, we need to discuss this river, its highly variable flow and unpredictable nature, and the numerous attempts by Native Americans and European settlers to use its waters. John Wesley Powell, one of the great government scientists of the nineteenth century, tried and failed to impose a framework on its water development. We also recount the little-known story of Lake Cahuilla and a disaster that led to a nationwide call for flow regulation on the Colorado River.

Human Use of Water in the Colorado River Basin

Archaic people living along the Santa Cruz River where it flowed through the present-day city limits of Tucson, Arizona, constructed the first known irrigation canals in the Colorado River basin—indeed, in all of North America—between two thousand and three thousand years ago. About one thousand years later, the Hohokam living where Phoenix is today began to build an extensive canal system, transporting water via hand-dug, gravity-fed canals up to ten miles from the Salt River. Most archaeologists believe that the collapse of Hohokam society was due in part to unpredictable river flows, in addition to population pressures, political turmoil, and environmental degradation several centuries before European settlers arrived on the scene.[5]

Elsewhere in the Colorado River basin, indigenous people constructed a variety of irrigation works, ranging from simple to complex. Along many watercourses, including the Colorado River itself,

---

5. M. Kyle Woodson and Gary Huckleberry, "Prehistoric Canal Irrigation in Arizona," in *Prehistoric Water Utilization and Technology in Arizona*, ed. Michael S. Foster, M. Kyle Woodson, and Gary Huckleberry (Phoenix, AZ: SWCA Inc., Environmental Consultants, and State Historic Preservation Office, 2002), 105.

farmers timed their plantings with natural flood cycles. Native Americans learned to use the river in spite of its quirks. In the Mojave Valley, near present-day Needles, California, the Mojave Indians had small-scale farms on the flood plains, seasonally growing a variety of crops from hand-diverted river water.[6] They and other groups, notably the Quechan and Cucapá who lived along the river from Yuma into the delta, harvested mesquite beans and grass seed produced by the abundant water. People harvested wood from trees and large shrubs for use in everything from arrow shafts to sandals to clothing to houses.[7] Those who followed had higher expectations for this hyperarid region.

The first Europeans to visit the Colorado River basin were Spaniards, who first arrived in 1540 and then made several passes through in the eighteenth century.[8] As European settlers accumulated in the region, they used the preexisting Native American canals, often augmenting and improving upon them.[9] Members of the Church of Jesus Christ of Latter-day Saints (a.k.a. the Mormons) arrived in northern Utah in 1847 under the leadership of Brigham Young. They immediately began irrigating their new farmlands. The irrigation systems that the Mormons perfected in northern Utah would become a model for future irrigation efforts in the American West.[10] The Mormons regulated rather minuscule watercourses compared to the Colorado River, which loomed as the prize for more ambitious irrigators.

Larger trends in American thinking promoted settlement of the West and plans for large-scale water diversions. In the mid-to-late nineteenth century, one of the most cherished concepts for American settlers was that of Manifest Destiny, the divine right to expand American borders westward to the Pacific coast.[11] One of

---

6. F. Kunkel, *The Deposits of the Colorado River on the Fort Mojave Indian Reservation in California, 1850–1969* (Menlo Park, CA: U.S. Geological Survey Open-File Report, 1970).
7. Anita Alvarez de Williams, *Travelers Among the Cucapa* (Los Angeles: Dawson's Book Shop, 1975), 24, 93–96, 129.
8. Charles K. Fox, "The Colorado Delta, A Discussion of the Spanish Explorations and Maps, the Colorado Silt Load, and its Seismic Effect on the Southwest" (unpublished manuscript, Los Angeles, 1936).
9. See Michael C. Meyer, *Water in the Hispanic Southwest: A Social and Legal History 1550–1850* (Tucson: University of Arizona Press, 1984), for more information about early Hispanic irrigation efforts.
10. Marc Reisner, *Cadillac Desert: The American West and Its Disappearing Water* (New York: Penguin Books, 1986); George Thomas, *The Development of Institutions Under Irrigation with Special Reference to Early Utah Conditions* (New York: Macmillan, 1920).
11. The term Manifest Destiny was coined by John L. O'Sullivan in his article advocating the annexation of Texas into the Union. John L. O'Sullivan, "Annexation," *The United States Magazine and Democratic Review* 17 (1845): 5.

the most vocal proponents of this idea was land promoter and politician William Gilpin. The nineteenth-century messiah of western land development, Gilpin believed that there was not a single square inch of land that could not be irrigated. He advocated the belief that "rain follows the plow," that is, the mere act of preparing land for crops would generate the precipitation needed to sustain them. It was a popular notion, one widely accepted across the country.[12] One person who did not accept this notion had extensive firsthand experience in the region, and he soon found that his science could be trumped by politics.

A Scientist Proposes an Irrigation Plan

John Wesley Powell, Civil War veteran and leader of daring Colorado River expeditions in 1869 and 1871–72, was a keen observer of landscapes. Modestly educated and possessing a strict military bearing, Powell earned his reputation as an intrepid explorer, first-rate scientist, and scientific visionary during his peerless career in the latter third of the nineteenth century.[13] While his reputation in the field of geology persists to this day, it was his contribution to hydrology, and more specifically, his plans for developing the desert landscapes of the United States, that got him deep into trouble and still generates a combination of admiration and debate.

Powell's 1878 *Report on the Lands of the Arid Region of the United States* had spelled out his revolutionary and controversial plan for land and water administration in the American West.[14] Powell advocated several critical concepts. He believed that water, like other natural resources, should be used to the greatest degree possible for the benefit of mankind; if it was necessary to drain a riverbed, so be it because "for the great purposes of irrigation ... the water has no value in its natural channel."[15] Second, water users, not governments, should be in charge of water management and control, which remained possible in the largely undeveloped West. Finally,

---

12. Thomas L. Karnes, *William Gilpin: Western Nationalist* (Austin: University of Texas Press, 1970); Stegner, *Beyond the Hundredth Meridian: John Wesley Powell and the Second Opening of the West* (Boston: Houghton Mifflin, 1962).
13. We relied on three major biographies written about John Wesley Powell: William C. Darrah, *Powell of the Colorado* (Princeton, NJ: Princeton University Press, 1951); Wallace Stegner, *Beyond the Hundredth Meridian*; and Donald Worster, *A River Running West: The Life of John Wesley Powell* (New York: Oxford University Press, 2001).
14. Powell, *The Arid Lands*.
15. Ibid., 54.

acknowledging that water, not land, was the limiting reagent in the agricultural formula for arid lands, Powell proposed land division based on water availability rather than neat blocks of a particular size, a system established in the eastern states in 1785.[16] Irrigation districts would be arranged in a hierarchy within a watershed, and the farmers living within any given district would cooperatively design, build, and maintain their own irrigation works within the confines of the overall plan for the watershed. When Powell became the second director of the nascent USGS in 1881, he finally had a chance to implement his ideas.

Powell's plan met with near-immediate and vehement opposition once it became known to land speculators and developers. He proposed that public domain lands were to be closed to entry until an irrigation survey could be completed, a delay greatly opposed by developers.[17] He alienated them further with his desire to ensure that homesteads actually went to and stayed with family farmers; under the extant system, wealthy landowners could easily enlist hired hands to homestead land on their behalf. In addition, Powell's water-district model overlooked established water users in certain regions, earning him additional enemies. States' rights advocates argued that the public domain should be turned over to the states, which would then build their own irrigation works; Powell's proposal reeked of federal intervention. Despite its bold vision, his plan was rejected by the early 1890s, which all but ended his professional career as a government scientist and bureaucrat.[18]

With the failure of Powell's comprehensive plan came an intensified piecemeal effort to divert surface water for irrigation throughout the Colorado River basin. The Mormons had demonstrated diversions on a small scale, and others had emulated or reactivated Native American canal systems. But the irrigation trophy remained

---

16. The Land Ordinance of 1785 established the Township and Range system for disposing of the public domain. Journals of the Continental Congress, 1774–1789, 28 (1785): 375–81, http://lcweb2.loc.gov/cgi-bin/query/r?ammem/hlaw:@field(DOCID+@lit(jc028100)) (accessed February 27, 2006).
17. Donald Worster, *Rivers of Empire* (New York: Oxford University Press, 1985), 131–42; Mary C. Rabbitt, *The United States Geological Survey, 1879–1989* (Washington, DC: U.S. Government Printing Office, U.S. Geological Survey Circular 1050, 1989), http://pubs.usgs.gov/circ/c1050/index.htm (accessed November 30, 2004); Worster, *A River Running West*, 474–75.
18. Mary C. Rabbitt, *Minerals, Lands, and Geology for the Common Defence and General Welfare*, vol. 1, *Before 1879* (Washington, DC, United States Government Printing Office, U.S. Geological Survey, 1979). While it is true that Powell remained as the director of the Bureau of American Ethnology until his death in 1902, his position and prestige were greatly diminished.

the mainstem Colorado River, particularly where it was bordered by broad fertile lands along the border between Arizona and California.

Challenges of a Free-Flowing River

Draining 256,700 square miles, the Colorado River is the fifth-largest drainage basin in the continental United States, but in terms of the amount of water it produces, it is the twenty-fifth largest. The explanation for this discrepancy centers on geography and climate; while the headwaters receive a prodigious snowfall, the average precipitation over the entire drainage basin is a mere eight inches per year. The snowmelt produces an annual runoff peak in May or June and sustains the perennial flow that occurs through the remainder of the year. More than 25 million people—this number is growing—depend on that water.[19]

In geologic terms, the Colorado River basin—from headwaters to sea—is relatively young, no more than 5 million years or so.[20] Before that time, details are limited, but the continuous river appears to have evolved from a chain of lakes that extended from northern Utah to northern Arizona and perhaps within the lower basin as well. Some believe that the Colorado River was a major tributary of the Rio Grande; others think it could have ended in a closed basin, much like the Mojave River in California.[21] Emerging geologic evidence downstream from Hoover Dam may ultimately yield more precise information on the question of when those lakes drained to form a continuous Colorado River.

We know that in the Pleistocene, 2 million to eleven thousand years ago, glaciers periodically formed and dissipated in the headwaters. We can speculate that the spectacular canyons that the river passes through on the Colorado Plateau were carved during the glacial meltwater floods issuing from the Rocky, Wind River, and Uinta

---

19. Colorado River Water Users Association Web site, http://www.crwua.org/colorado_river.html (accessed March 13, 2006).
20. P. Kyle House, Philip A. Pearthree, Keith A. Howard, John W. Bell, Michael E. Perkins, James E. Faulds, Amy L. Brock, "Birth of the Lower Colorado River—Stratigraphic and Geomorphic Evidence for its Inception near the Conjunction of Nevada, Arizona, and California," in *Interior Western United States: Geological Society of America Field Guide 6*, ed. Joel L. Pederson and Carol M. Dehler (Boulder, CO: Geological Society of America, 2005), 357–87.
21. A large divergence of opinion on the origin of the Colorado River and downcutting of Grand Canyon is given in *Colorado River Origin and Evolution*, ed. R. A. Young and E. E. Spamer (Grand Canyon, Arizona: Grand Canyon Natural History Association, Monograph 12, 2001).

mountains. All we really know about prehistoric floods comes from an archaeological site two river miles downstream from Lee's Ferry, where radiocarbon-dated flood deposits suggest a peak discharge of 500,000 cubic feet per second (ft^3/s) occurred 1,200 to 1,600 years ago.[22] In all likelihood, the magnitude of this flood was small in comparison with what might have occurred in the late Pleistocene.

The settlers of the Colorado River basin quickly had to deal with the large floods that the river produced. The 1862 flood destroyed several towns in the Virgin River basin as well as Yuma;[23] later in the twentieth century, hydrologists estimated the peak discharge of this flood to be 400,000 ft^3/s downstream from Needles, California. A flood of unknown size, generated from both the lower Colorado and Gila rivers, inundated about three-quarters of Yuma in January 1874.[24] In 1884, several floods swept down the river, damaging towns and irrigation works in the upper basin. On the basis of a remembered tree branch where a rabbit perched above the floodwaters at Lee's Ferry,[25] hydrologists estimated the peak discharge of this flood to be 300,000 ft^3/s. Not to be outdone, the Gila River, which drains about two-thirds of Arizona, produced its own 300,000 ft^3/s flood in the winter of 1891 that valiantly tried to wash Yuma from the map.

The river had the upper hand along its channel. One way to visualize its impact is to imagine what a firehose, fully pressurized and turned on, does when released onto the ground. Where unconfined, as on an open field, the hose swings wildly and unpredictably. If laid out in a narrow alley, the hose continues to wriggle out of control, only within a more confined place. In its canyons, the Colorado River was like that firehose in an alley—it remained in a narrow corridor, rising up and down with the annual flood but changing position within a narrow band. But in the wide lower canyons, and especially in its delta, the river channel shifted over many miles of alluvial bottomland. Diverting water for irrigation, even determining where to build a town, becomes problematic when the river might shift through downtown or miles away from the canal intake.

---

22. J. E. O'Connor, L. L. Ely, E. E. Wohl, L. E. Stevens, T. S. Melis, V. S. Kale, and B. R. Baker, "A 4500-year record of large floods on the Colorado River in the Grand Canyon, Arizona," *Journal of Geology* 102 (1994): 1–9.
23. R. W. Durrenberger and R. S. Ingram, *Major Storms and Floods in Arizona, 1862–1977* (Tempe, AZ: Office of the State Climatologist, Climatological Publications, Precipitation Series No. 4, 1978); W. N. Engstrom, "The California storm of January 1862," *Quaternary Research* 46 (1996): 141–48.
24. Durrenberger and Ingram, *Major Storms and Floods*, 2.
25. Reilly, *Lee's Ferry*, 99, 290.

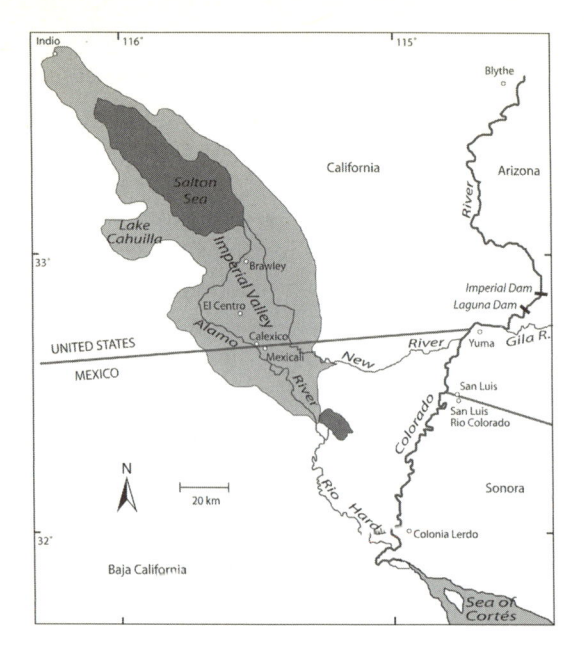

The lower Colorado River and the Salton Sink.

The Colorado River had an even bigger trick awaiting the settlers in the lower basin. It flows to the Sea of Cortés through its delta, a large area of low relief. The Sea of Cortés formed from the rending of a continent by large faults, and the delta is the accumulated detritus from a large part of North America, a part known for its substantial sediment production. The faults, which include the notorious San Andreas Fault responsible for the 1906 earthquake in San Francisco, continue to shift and twist the surface. The Salton Sink literally has sunk, with a base 287 feet below sea level, under the influence of those faults and their local disruption of the Earth's surface here. Approximately twenty thousand feet of sediment has accumulated in this closed basin.[26]

William Phipps Blake, a geologist with the Army's Corps of Topographic Engineers, came to the lower Colorado region in the mid-1850s as part of a railroad route survey team.[27] He observed

26. S. Jenning and G. R. Thompson, "Diagenesis of Plio-Pleistocene sediments of the Colorado River delta, southern California," *Journal of Sedimentary Petrology* 56 (1986): 89–98; T. Parsons and J. McCarthy, "Crustal and upper mantle velocity structure of the Salton Trough, southeast California," *Tectonics* 15 (1996): 456–71.
27. William P. Blake, "Sketch of the Region at the Head of the Gulf of California: A Review and History," in *The Imperial Valley and the Salton Sink* by Henry T. Cory (San Francisco: John J. Newbegin, 1915), 1–35; Joseph E. Stevens, *Hoover Dam: An American Adventure*

The western side of the Salton Sink in 1905. Lake Cahuilla shorelines are painted in calcium-carbonate deposits along the mountain fronts at left.

W. C. Mendenhall 512, courtesy of the U.S. Geological Survey Photographic Library.

---

that much of what is now Imperial County, California, was once an arm of the Sea of Cortés, and the river's mouth would have been near present-day Yuma. As the muddy river's silt load dropped, it built a dam, closing off the northern tip of the sea. The remnant lake, which Blake named Lake Cahuilla, eventually evaporated, leaving behind salts capping its barren sediments.

Those sediments came mostly from the Colorado River. Periodically, just downstream from Yuma, the channel would swing far enough west to break through the low divide and flow northwest into the Salton Sink, disgorging most, if not all, of its flow into the closed basin.[28] Lake Cahuilla filled, usually to levels well above sea level, before the river switched back to its delta. As its waters rapidly evaporated under the torrid heat of the lower Sonoran Desert, the

---

(Norman: University of Oklahoma Press, 1988), 9.

28. P. C. Van de Kamp, "Holocene continental sedimentation in the Salton Basin, California: A reconnaissance," *Geological Society of America Bulletin* 84 (1973): 827–48.

lake receded until it formed a salt pan, not unlike playas elsewhere in the California deserts. Native Americans took advantage of this receding lake by trapping fish, stocked from the river as the lake filled.[29] The Salton Sink periodically filled during the Holocene, forming Lake Cahuilla at least four times between A.D. 700 and 1580.[30] Sedimentological and archaeological evidence suggests that this lake may have been present for as much as three-quarters of the last 1,700 years. Blake recognized that occasionally water still flowed into the Salton Sink, temporarily converting it back into a lake. The same favorable slopes that made irrigation in the Colorado Desert so appealing also made it tricky: it would take very little to convince the entire river to transfer its flow from sea to sink.[31]

Some thought it would be great if the river would flow into the Salton Sink. Dr. J. P. Widney's plans for the lower Colorado River were based on a twist on Gilpin's ideas. Starting in 1873, Widney spent a couple of years in Yuma and observed the remains of Lake Cahuilla firsthand.[32] He hypothesized that the drying of the lake caused drying of the surrounding desert, and if only the Colorado River were convinced to flow back into the Salton Sink, the climate would change to a more temperate regime compatible with self-sufficient agriculture. Evaporation from the lake would create lake-effect increases in rainfall downwind, or so his theory went. Fortunately, this theory, bereft of scientific foundation, went untested.

Widney's notions aside, settlers were still trying to grow crops in the Colorado Desert. They believed all they needed was water to make the desert bloom, and all the water they needed was in the Colorado River. The river had a troublesome history, however. Spanish expeditions in the seventeenth century had seen a lake in the Salton Sink with a water elevation at about sea level; this lake reportedly began to recede by A.D. 1640.[33] Smaller breaches leading to smaller lakes occurred in 1828, 1840, 1849, 1862, 1867, and 1884.[34] In 1891, the river cut through its banks and shifted part

---

29. Kenneth W. Gobalet, Thomas A. Wake, and Kalie L. Hardin, "Archaeological record of native fishes of the lower Colorado River: How to identify their remains," *Western North American Naturalist* 65 (2005): 335–44.
30. M. R. Waters, "Late Holocene lacustrine chronology and archaeology of ancient Lake Cahuilla, California," *Quaternary Research* 19 (1983): 373–87.
31. Stevens, *Hoover Dam*; William deBuys and Joan Myers, *Salt Dreams: Land and Water in Low-Down California* (Albuquerque: University of New Mexico Press, 1999).
32. George Wharton James, *The Wonders of the Colorado Desert* (Boston: Little, Brown, 1906), 2:271–76.
33. Fox, *The Colorado Delta*.
34. MacDougal, "The delta of the Rio Colorado;" Godfrey Sykes, "The delta and estuary

of its flow into the Sink for a short period. The flow was sufficient to allow small boats to pass from Yuma into the incipient lake.[35] The banks healed naturally after the annual peak, however, and the river resumed its course into the Sea of Cortés. Years of drought followed, and the suddenly docile river convinced developers that the river could be used to irrigate crops. After an initial failure, large-scale plans were made, and set into action a series of events that led to a major federal investment in water storage and delivery infrastructure of water in the western United States.[36]

The Early Attempts at Diversion

It must take a different kind of eyesight to see a barren desert and visualize productive farmland. But people have possessed such vision for millennia. One need look no further than the Tigris and Euphrates rivers in the Fertile Crescent (in present-day Iraq) where agriculture was first developed.[37] Perhaps the pinnacle of Old World agricultural development in a desert environment is the Nile through Egypt, where the river is fully developed for agricultural irrigation. Like the Nile, the Colorado River deposits sediments that are both fertile and appropriate for farming, in this case, in the Salton Sink. Looking a little beyond the horizon, the lower Gila River also possessed fertile farmland, but, if it were conceivably possible, the Gila was a more fickle river than the Colorado as a source of irrigation water.

The first attempt at large-scale diversion along the mainstem Colorado occurred not far upstream of Yuma in 1857.[38] This ambitious irrigation system consisted of a canal with viaducts over dry washes, and it was a total failure. Smaller brush dams, thrown up during low flow periods, quickly were swept away during the annual flood. The settlers did not want to irrigate by hand or wait for the annual flood to subside, as the Native Americans did, and larger schemes were soon hatched.

---

of the Colorado River," *Geographical Review* 16 (1926): 232–55; Godfrey Sykes, *The Colorado Delta* (Washington, DC: The Carnegie Institution of Washington and the American Geographical Society, 1937); W. G. Hoyt and W. B. Langbein, *Floods* (Princeton, NJ: Princeton University Press, 1955).

35. Sykes, *The Colorado Delta*, 40.
36. Readers should see Reisner, *Cadillac Desert*, for a more complete, if biased, history of water development in the West. Here, we concentrate on some specific issues related to development of the lower Colorado River.
37. Jared Diamond, *Guns, Germs, and Steel: The Fates of Human Societies* (New York: W.W. Norton, 1998), 134.
38. Douglas D. Martin, *Yuma Crossing* (Albuquerque: University of New Mexico Press, 1954).

Oliver Meredith Wozencraft, a physician by training, had succumbed to the siren song of the California Gold Rush in 1849. Crossing the Colorado Desert west of Yuma in May, he became delirious and was gripped by a mirage of lush greenery. William Blake's report on the land's potential—if one ignored those pesky caveats about Lake Cahuilla—provided him with the impetus to transform vision into quest. For many years in the late nineteenth century, he lobbied in California and Washington for title to ten million acres of desert and the right to irrigate it with Colorado River water, transported through a canal just north of the border. Wozencraft died in 1887 at the age of seventy-three, his hopes unrealized. But his idea was too enticing to escape the notice of others.[39]

At about the same time Wozencraft was advocating his plan, financier Thomas Henry Blythe, who had made his fortune in San Francisco real estate, was devising a development scheme for lands farther upstream along the Colorado. San Diegan O. P. Calloway, who was interested in irrigating lands in what is now known as the Palo Verde Valley, initially approached Blythe in the 1870s. When members of one of the river tribes killed Calloway, Blythe continued the efforts on his own, with plans to bring into cultivation some 175,000 acres for "An Empire on the Colorado" ninety miles north of Yuma. Blythe also entered into a partnership with Mexican General Guillermo Andrade to develop water and lands south of the border. At a ranch north of Colonia Lerdo in the delta, their company constructed a canal that was used to irrigate fruit trees and grain fields in 1883, and they had plans for a larger canal in the Mexicali Valley, with settlements to be populated with European immigrants. When Blythe died suddenly in April 1883, both of his projects faltered, although his name was given to a California town that developed in the Palo Verde Valley.[40]

First Disaster, Then an Emerging Plan

Engineer Charles Robinson Rockwood first came to investigate irrigation possibilities in the delta region in 1892, and bringing

---

39. Stevens, *Hoover Dam*, 8–10; deBuys and Myers, *Salt Dreams*, 76–77; George Kennan, *The Salton Sea: An Account of Harriman's Fight with the Colorado River* (New York: Macmillan, 1917), 16–18.
40. William O. Hendricks, "Guillermo Andrade and Land Development on the Mexican Colorado River Delta 1874–1905" (doctoral diss., University of Southern California, 1967), 106–109; William O. Hendricks, "Developing San Diego's Desert Empire," *Journal of San Diego History* 17 (Summer 1971), http://www.sandiegohistory.org/journal/71summer/index.htm (accessed March 24, 2005).

Colorado River water to the Colorado Desert quickly became his obsession. After a failed attempt or two, Rockwood created the California Development Company (CDC), and found a backer, George Chaffey, a wealthy engineer with experience in irrigation projects in Australia. The CDC tossed aside the desolate moniker of Colorado Desert, replacing it with the more tantalizing Imperial Valley to attract investment.[41]

Chaffey thought the best way to transfer Colorado River water to the Imperial Valley was to link a canal with the channel bed of the Alamo River, which meandered through Mexico before winding northwards into California. He and Rockwood likely did not realize or appreciate that they had identified the same topographic gradients that the river used when it filled Lake Cahuilla, as well as the same channel. All that was needed was to breach the levees, built up by the river's prodigious sediment load, with a headgate. A short canal segment connected with the natural distributary channels—multiple channels going to the same destination—that flowed west and northwest into the Imperial Valley and northern Mexico. In May 1901, the CDC opened a hastily constructed headgate at Hanlon Heading, just north of the U.S.–Mexico border, and began to deliver the first irrigation waters to farmers in the Imperial Valley.

Difficulties quickly plagued the fledgling company. The first came in the form of an unfavorable and widely-distributed 1902 report from the Department of Agriculture concerning the excessively high alkalinity in the soils of the Imperial Valley.[42] A far more devastating blow followed in the same year with the creation of the U.S. Reclamation Service; among other duties, Reclamation would manage the nation's waters, particularly those associated with navigable rivers, such as the Colorado.[43] Rockwood's troubles were

---

41. Several publications include detailed accounts of Rockwood and the 1905 breach that led to the filling of the Salton Sea. We drew upon Cory, *The Imperial Valley and the Salton Sink*; Kennan, *The Salton Sea*; Norris Hundley, "The Politics of Reclamation: California, the Federal Government, and the Origins of the Boulder Canyon Act—A Second Look," *California Historical Quarterly* 52 (Winter 1973): 299–305; Alton Duke, *When the Colorado River Quit the Ocean* (Yuma, Arizona: Southwest Printers, 1974); Worster, *Rivers of Empire*; deBuys and Myers, *Salt Dreams*; and Stevens, *Hoover Dam*.
42. Thomas H. Means and J. Garnett Holmes, *Soil Survey Around Imperial, Cal.* (Washington, DC: U.S. Department of Agriculture, Bureau of Soils, Circular No. 9, January 10, 1902).
43. Institute for Government Research, *The U.S. Reclamation Service: Its History, Activities and Organization* (New York: D. Appleton, Service Monographs of the United States Government No. 2, 1919), 16–23. This volume contains a wealth of information about the early years of the Bureau of Reclamation.

compounded when the inadequately engineered Hanlon headgate silted in sufficiently to render it useless, leaving angry water users high and dry. Because the United States government would not grant permission to construct a new headgate on American soil, the CDC turned to Mexico. Mexican officials agreed to the plan, but only if half the diverted water went to the Mexicali Valley, south of the border. The CDC was desperate, so it accepted the Mexican offer. Desperation rarely breeds care, and in this case, the new Mexican cut, built in 1904, completely lacked a controlling headgate. To keep the intake clear of sediments, the CDC increased the gradient of the channel flowing westward.

Siltation, high soil alkalinity, unstable headgates, the Mexican government, and the U.S. Reclamation Service aside, the CDC was successfully promoting agricultural development of the Imperial Valley. The *Los Angeles Times* gushed, "vast areas of the great Colorado Desert are being transformed into smiling agricultural plains."[44] That the *Times* would be so effusive in its praise did not come as much of a surprise. Publisher Harrison Gray Otis and his son-in-law, Harry Chandler, who was vice-president and general manager of the paper, presided over a group of investors who, in 1904, had purchased over 860,000 acres of irrigable land in the Mexicali Valley from Guillermo Andrade.[45] Otis and Chandler, wealthy insiders in the Los Angeles scene, were concurrently involved in another get-rich-from-imported-water scheme. They were part of a syndicate that acquired a huge chunk of the San Fernando Valley east of Los Angeles, where the city planned to store the water it planned to import from the Owens River, nearly 250 miles away.[46]

From 1891 through 1904, drought prevailed in the region, and the Colorado River responded with relatively low flows. The relative calm of the river enabled the cash-strapped CDC developers to put off installing controls on the troublesome canal intakes. Yet without headgates, the flow of water down the canal could not be shut off, even if the CDC wanted to do so. In addition, the CDC protected the banks around the intake structure for only short distances both upstream and downstream. Combined with the straight, relatively steep channel leading away from the headgates, conditions were perfect for another chapter in the history of Lake Cahuilla.

---

44. "How the Great Colorado Desert is Being Made to Blossom," *Los Angeles Times*, January 1, 1905 (accessed December 27, 2004).
45. deBuys and Myers, *Salt Dreams*, 142. An expose of the Otis-Chandler dynasty is available in William G. Bonelli, *Billion Dollar Blackjack* (Beverly Hills, CA: Civic Research Press, 1954).
46. Reisner, *Cadillac Desert*, 75–77.

The floods of 1905, now notorious in regional history, were actually not that large. The first one, on March 20, peaked at 112,000 ft^3/s.[47] This flood was 50–110% larger than the peaks of the previous two years, and the CDC was unprepared. The floodwaters cut through the unregulated canal intake, flowed full force to the west, then turned north. The developers knew about the existence of the New River; it appeared on maps made several years earlier.[48] What they may not have realized was that the combination of the New and Alamo rivers, which they had envisioned as irrigation canals, were actually the ideal pathways for water to flow into the Salton Sink. Within a year, and despite the frantic attempts to plug the breach, the Colorado River flowed fully towards the northwest, abandoning its delta in favor of the Sink.[49]

Rockwood had meanwhile forced Chaffey out, and with an eye to the unfolding disaster, he needed new backers. While the annual flood was still rising in June, Rockwood negotiated a loan from Edward H. Harriman, president of the Southern Pacific Railroad and a member, along with Otis and Chandler, of the San Fernando Valley syndicate. Harriman, as part of the agreement, restructured the CDC and installed one of his executives, an engineer named Epes Randolph, as company president. Harriman was not fully aware of what he had acquired, but Randolph soon filled him in: a roiling flood was eating away at the Mexican banks of the river, converting the once irrigation canal fully into the Colorado River. They tried a remedy of a second, regulated headgate, but the November flood, which peaked at 109,000 ft^3/s on the thirtieth, ripped it away, and most of the Colorado River continued to flow unimpeded to the northwest.

Nothing the engineers threw at the banks, including vast sums of money, seemed to stem the flow. In the spring of 1906, Rockwood either resigned or was fired. H. T. Cory, an engineer from within the Southern Pacific ranks, replaced him. Cory's task was formidable, to say the least. He built a branch railroad to the breached banks, then constructed a rock dam atop a brush mattress foundation that

---

47. J. L. Patterson and W. P. Somers, *Magnitude and Frequency of Floods in the United States. Part 9. Colorado River Basin* (Washington, DC: U.S. Government Printing Office, U.S. Geological Survey Water-Supply Paper 1683, 1966), 464.
48. T. H. Silsbee, "Map of That Part of the Colorado Desert in California, United States and Lower California, Mexico, Known as the New River Country" (San Diego: California Land Development Company, unpublished blue-line map, scale approximately 1:178,000, 1900).
49. Stevens, *Hoover Dam*.

prevented the rocks from sinking into the seemingly bottomless river mud. When this failed, Cory built a trestle over the break, and using every available Southern Pacific railroad car, ordered enormous quantities of rock dumped into the river. Finally, after spending millions of dollars, on February 10, 1907, Cory had the Colorado River flowing again to the Sea of Cortés.

The "Menace" Sustained

Despite Cory's pyrrhic victory—the canal system was now a mess, the Imperial Valley farms were dry, and Harriman was pleading for federal relief of his enormous debt—the control of the river was far from being over. Between 1905 and 1907, Lake Cahuilla did not reach its previous high-lake levels, and perhaps to separate the human-created lake from the natural one, they renamed the water body the Salton Sea. The fledgling agricultural industry suffered millions of dollars in damages, not to mention the loss of its irrigation lifeline. Now the warnings were fully realized: the fast-moving headcuts, or waterfalls, developed in the alluvial fill of the Imperial Valley and quickly migrated upstream, leaving behind deeply incised channels. Luckily, those channels never reached the Colorado River; had they done so, the river might have downcut into its bed upstream from Yuma, stranding irrigation headgates and making its return to the delta even more difficult. Now two flow regulation features were required: a headgate that could be controlled to ward off flood flows, and a weir that would prevent future downcutting.

Laguna Dam, a concrete flow-diversion structure completed in 1909, was expected to meet both of these requirements. The citizens of Yuma had enthusiastically celebrated its authorization in January 1905.[50] That the "dam" would not provide complete flood control was dramatically emphasized during the dedication ceremony on March 31, 1909, when an "unexpected rise" in the river tore away thirty feet of a cement wing protecting an intake canal. The "miniature Niagara" muddily cascading over the dam stole both thunder and voices of the undoubtedly far less interesting dedication speakers, but the celebrants were undaunted in their appreciation of the new structure, which, broken wing aside, ultimately performed its task.[51]

50. "Yuma People Feel Jubilant," *Los Angeles Times*, January 10, 1905 (accessed December 27, 2004).
51. "Christen in Flood," *Los Angeles Times*, April 1, 1909 (accessed December 27, 2004).

In 1909, the Colorado once again jumped into another old channel, the Abejos (Bee) River, which drained into Volcano Lake south of the California–Mexico border. Fearful of a Salton Sea sequel, Congress appropriated $1 million in 1910 to remedy the situation, hiring noted St. Louis engineer John Ockerson to direct the work. Ockerson quickly constructed a weir, but he reported "that the Imperial Valley would never be safe from the menace of western diversions due to flood-waters until ample works were constructed to confine such flood-waters to narrow limits along the river proper."[52] He had in mind more levees—miles and miles of levees—only this time in Mexico. Then, as Cory noted, "While operations were in progress, the Revolution in Mexico began, and resulted in the abdication of President Diaz. On February 21st, the Revolutionists captured Algodones, and took possession of a work train for half a day. On April 16th a large body of Mexican Federal troops arrived at the break and remained guarding the work from interruption until May 10th."[53]

Through the turmoil south of the border, farmlands to the north were dry, and farmers began to search for solutions. One was rooted in California's 1887 Irrigation District Act, which allowed communities to organize into official districts that would cooperatively construct and manage irrigation projects. The act was fused as a result of California's complicated rules covering who had the right to a river's water: those who held land along its banks (the "riparianists," who were often ranchers running stock) as opposed to those who had first used the water but whose land might be at a distance from a river (the "appropriationists," after the doctrine of prior appropriation, who were often farmers). In California, riparianists prevailed, putting those without riverside acreage at a great disadvantage. By forming irrigation districts, California farmers could claim the necessary lands on which to build the irrigation works they desired.[54]

In 1911, community leaders in Imperial County, California argued persuasively on behalf of an irrigation district, and voters created the Imperial Irrigation District (IID).[55] One of the IID's first

---
52. Cory, *The Imperial Valley and the Salton Sink*, 1439.
53. Ibid., 1443.
54. Worster, *Rivers of Empire*, 107–9. For a thorough discussion of riparianism and prior appropriation, see Joseph L. Sax, Robert H. Abrams, and Barton H. Thompson Jr., *Legal Control of Water Resources: Cases and Materials*, 2nd ed., American Casebook Series (St. Paul, MN: West Publishing, 1991), 37–317.
55. Norris Hundley, *Water and the West: The Colorado River Compact and the Politics of Water*

goals was to acquire the struggling remnants of the CDC, which, due to legal complications, took five years. Imperial Valley farmers were acutely aware of their absolute dependency on a reliable flow of water from the Colorado River and had several specific concerns. One was the river itself; its predilections seemed to exclude dependable irrigation and were viewed as a threat to both land and life. That all water flowed through Mexico prior to reaching them thoroughly annoyed Imperial Valley residents, who sometimes saw the battered body of a Mexican revolutionary war victim or fetid animal carcasses drifting along in what, after all, was their domestic water supply. Further, Mexican landholders were not only entitled to half of the canal's flow, they had first crack at it. As agriculture in the Mexicali Valley rapidly increased, it seemed entirely plausible that the water available to Imperial Valley would dwindle to nothing. The IID's solution was to construct a canal situated entirely north of the border—an "all-American" canal.[56]

The IID faced strong opposition from Harry Chandler's syndicate (Harrison Otis died in 1917), as the canal would have devastating effects on Chandler's extensive Mexicali Valley farms. The multi-million dollar canal price tag, prohibitive for the IID constituency, was also daunting. Most importantly, the IID needed to control those Colorado River floods, a task far beyond its fiscal and territorial capacity. In spite of their reluctance to work with the federal government, members sent their legal counsel, Phil Swing, to visit with the Secretary of the Interior, who agreed to a canal survey, the costs to be jointly borne by the government and the IID.[57]

Although Cory healed its banks with quarried rock, the lower Colorado River continued to wreak havoc. The drought ended in 1905, and the subsequent wet period greatly raised flow in the Colorado River. Large floods occurred in the upper basin, notably in September 1909 and October 1911.[58] In January 1916, the Gila River swelled to over 200,000 ft^3/s, deluging Yuma and backing the Colorado River up to Laguna Dam. Yuma mayor Charles Moore reportedly "died of heart failure, caused by the excitement

---

in the American West (Berkeley: University of California Press, 1975), 28–29.
56. Worster, *Rivers of Empire*, 201–3; deBuys and Myers, *Salt Dreams*, 128–29, 158–59; Hundley, *Water and the West*, 86; Hundley, "The Politics of Reclamation," 305–6.
57. Hundley, *Water and the West*, 42; Beverly Moeller, *Phil Swing and Boulder Dam* (Berkeley: University of California Press, 1971), 15–16; Hundley, "The Politics of Reclamation," 310–12.
58. Ralph R. Woolley, *Cloudburst Floods in Utah* (Washington, DC: U.S. Government Printing Office, 1946).

in trying to restore order."[59] Floods in June 1920 and June 1921, which peaked at 190,000 and 188,000 ft^3/s at Yuma, respectively, were the largest of the twentieth century.

Heart-stopping floods made the papers, and the cry to subjugate the "tortuous and troublesome river"[60] grew in intensity, within and outside the Imperial Valley. Calls for flood control became howls for federal government intervention, a once-despised notion. The 1905 breach was filled, but threats to the limited irrigation works on the river continued to occur and had to be dealt with on an incident-by-incident basis. Of more importance to would-be irrigators, flow was low during the months when irrigation demand would be highest. What was needed was flow regulation to smooth out that large intra-annual variation. The "natural menace" of the Colorado River needed the control offered by a large dam and a regional plan.

---

59. "Rebuild Yuma People Demand," *Los Angeles Times*, January 24, 1916 (accessed December 27, 2004).
60. "Struggling with a Tortuous and Troublesome River," *Los Angeles Times*, June 18, 1911 (accessed December 27, 2004).

2

Where Should the Dams Be?

Politics, the Colorado River Compact, and the Geological Survey's Role

The 1905 disaster and the subsequent need to reestablish the irrigation network in the Imperial Valley focused national attention on the lower Colorado River. The ongoing flood troubles held that attention, even through the turbulent years of World War I. It greatly helped that the *Los Angeles Times*, a hardly unbiased newspaper given its owners' land holdings in Mexico, continued to beat the drum for river management and irrigation projects. Politicians could not ignore the enormous agricultural potential of the region, not to mention the large profits associated with fully implemented water development. The coastal areas of Southern California—rapidly developing with towns and agricultural areas—needed water, and having developed all local sources, politicians and water managers began to look north, northeast, and east to meet their growing thirst. The Colorado River looked to be the perfect source, and Southern Californians used the disaster to gain traction in their goal for regulating the "out-of-control" river.

It is one thing to advocate flow regulation, and quite another to implement it. The California Development Company had ceased to exist, and the Imperial Irrigation District did not have the resources to fund dam investigations. In 1910, Arizona wasn't even a state, and California was more focused on developing its

Central Valley resources. It was clear that sooner or later the federal government would become involved in solving the problem of the lower Colorado River. This involvement tested the boundaries between science and politics within two closely related bureaus of the Department of the Interior. The U.S. Geological Survey (USGS) came first, with its roots firmly in the nineteenth century, followed by the U.S. Reclamation Service, a USGS by-product borne of John Wesley Powell's ideas as well as the nation's need to sustain agriculture in its arid regions.

The World's Oldest Government Science Agency

On March 3, 1879, President Rutherford B. Hayes signed a sundry civil expenses bill, which included just over a page of legislation establishing a United States Geological Survey.[1] Such brevity seems out of proportion with the significance: USGS became the world's first government-funded agency devoted to science. The first director was Clarence R. King, who had recently led the Army Corps of Engineers' Geological Exploration of the Fortieth Parallel; he was followed by John Wesley Powell in 1881. It was Powell who shaped the agency into the rough framework needed to address water-resources issues in the western United States.[2]

Throughout most of its long history, USGS has rested on a foundation of three disciplines: topographic mapping, geology and geologic mapping, and water-resource assessments. These disciplines, combined with engineering design, are essential to the location, sizing, and construction of dams. Topography helps determine the potential size of reservoirs, geology helps to determine the seismic stability of a dam site as well as the stability of the host bedrock, and hydrology determines how much water is to be expected and at what rate of flow. With the emergence of a grandiose plan for water development in the Colorado River basin, the federal government needed the expertise provided by its then-primary science agency. That expertise derives from the vision of Powell and the wisdom of the third director of USGS, Charles D. Walcott.[3]

---

1. Geological Survey subsection of *An Act Making Appropriations for Sundry Civil Expenses of the Government for the Fiscal Year Ending June Thirtieth, Eighteen Hundred and Eighty, and for Other Purposes*, 45th Cong., 3d sess., *U.S. Statutes at Large* (1879), 20:394–95.
2. Rabbitt, *Minerals, Lands, and Geology*, vol. 1; Mary C. Rabbitt, *A Brief History of the U.S. Geological Survey* (Reston, VA: U.S. Geological Survey, 1979); Worster, *River Running West*.
3. Biographies of Walcott are available in N. H. Darton, "Memorial of Charles Doolittle Walcott," *Bulletin of the Geological Society of America* 39 (1928): 80–116; and Ellis L.

Hydrology and the USGS

Implementing his report on the arid lands, Powell had invested USGS with a Hydrography Division, whose mission was water development and planning. The first phase of hydrologic science was to gather the basic data on river flow. Streamflow gaging in the United States began with the War Department and its measurements of the Connecticut River in 1871. In 1878, California established a dozen gaging stations in its Central Valley, and, in the 1880s, Colorado measured streamflow from the east slope of the Rocky Mountains.[4] Powell was determined to make streamflow gaging a federal occupation as well, controlled by USGS.

In fall 1888, Powell hired Frederick Haynes Newell as the first employee of his Irrigation Survey.[5] Newell promptly followed Powell's instructions and created a camp for the purposes of learning hydrography, the term then used to describe the measurement of flow at gaging stations. The gaging station they installed on the Rio Grande at Embudo, New Mexico—the first USGS gaging station in the United States—became operational on January 1, 1889.[6] The first station producing lasting data in the Colorado River basin was installed on the Green River at Green River, Utah, in 1895. Although measurements were made near Grand Junction, Colorado, in the late nineteenth century, the first systematic flow measurements on the mainstem Colorado River were at Yuma, beginning in 1903. It was at this site, with its unstable bank and beds, that USGS hydrologists learned that they needed more stable measurement sites to assess water flows, and, funded by Congressional appropriations to assess national water resources, they subsequently fanned out to all accessible sites along the Colorado River in Arizona, Utah, and Colorado, seeking to identify any and all possible locations for gaging stations.[7]

Yochelson, *Charles Doolittle Walcott, Paleontologist* (Kent, Ohio: Kent State University Press, ca. 1998).

4. Robert Follansbee, *A History of the Water Resources Branch of the U.S. Geological Survey: Volume I, From Predecessor Surveys to June 30, 1919* (Washington, DC: U.S. Government Printing Office, 1994), 31–32.
5. Follansbee, *A History*, 1:28. For insights into Newell's role within the Reclamation Service and American irrigation, see also Donald C. Jackson, "Engineering in the Progressive Era: A New Look at Frederick Haynes Newell and the U.S. Reclamation Service," *Technology and Culture* 34 (July 1993): 539–74.
6. Kenneth L. Wahl, Wilbert O. Thomas, Jr., and Robert M. Hirsch, *Stream-Gaging Program of the U.S. Geological Survey* (Reston, VA: U.S. Geological Survey Circular 1123, 1995), http://pubs.usgs.gov/circ/circ1123/ (accessed February 20, 2006).
7. Follansbee, *A History*, 1:53–56.

Frederick Haynes Newell, first director of
the U.S. Reclamation Service.

Photographer unknown, Portraits 379, courtesy of the U.S. Geological Survey Photographic Library.

At the same time, planning for dams required at least rudimentary sizing of reservoirs. Because most of the potential dam sites were in largely inaccessible canyons, the topographic mapping begun by Powell in 1869 had insufficient resolution to determine the potential volume of a reservoir. USGS, tasked with topographic mapping of the United States, had a cadre of surveyors on its payroll who were highly skilled at making maps, particularly in the rugged river canyons of the West. They were engaged to make the more detailed topographic measurements needed to size reservoirs and plan the locations of potential dam sites.

With Powell's departure from USGS in 1894, the Hydrography Division became a political hot potato. Charles D. Walcott became director and inherited a fiscal mess: his budget was extremely tight, reined in by the combination of a recession and continued animosity from Congress about Powell's failed irrigation plan. Walcott told Newell to either find additional funds or resign. The

Arthur Powell Davis, second director of
the U.S. Reclamation Service.

Courtesy of the U.S. Bureau of Reclamation.

resourceful Newell successfully lobbied Congress for a modest allotment for streamflow gaging, thus keeping his job and the Hydrography Division. Powell had justified this division with his irrigation plan, but now a new mission was needed to thwart the political animosity.[8] That mission became an overall assessment of the nation's water resources, a mission that continues to this day.

The director of the Topographic Branch, Henry Gannett, quickly approached Newell with a personnel issue. As a money-saving gesture, Gannett wanted to reassign one of his topographers to Hydrography, and Newell chose Arthur Powell Davis.[9] Davis, whose mother was John Wesley Powell's sister, began his USGS career in 1882 as a topographic engineer and had expressed interest in hydrography "due to his belief in the important roles that were

---

8. Follansbee, *A History*, 1:49; Rabbitt, *Minerals, Lands, and Geology*, 2:233–39.
9. For further information about Davis, see the memoir by Charles A. Bissell and F. E. Weymouth, "Arthur Powell Davis, Past-President, Am. Soc. C.E.," *Transactions of the American Society of Civil Engineers* 100 (1935): 1582–91; and Hundley, "The Politics of Reclamation," 292–325.

destined to be played in the future by water and stream gaging."[10] It was a portentous reassignment.

While Davis and a handful of other employees established and operated gaging stations, Newell spent much of his time promoting hydrographic work with both the public and politicians. His efforts, aided by a changing political climate, were rewarded. Attendees of the 1895–98 Irrigation Congresses begged for the continuation of irrigation surveys under the Department of the Interior, having decided that the irrigation works they desired called for federal government participation. The collective cry of the states for their right to build and control irrigation works had also weakened, in part due to the failure of the 1894 Carey Act. The Carey Act granted public domain acreage to arid lands states that wished to reclaim it, but most states simply did not have the economic resources to construct massive public works. There was also the problem of rivers flowing between states: which state had the rights to such waters? Furthermore, the global depression of 1893 eliminated the prospect of private, international funding of irrigation projects.[11]

A group of seven men, which included Newell, formed the National Irrigation Association in 1899. Their specific goal was to lobby for federalized control and subsidy of irrigation projects. At the 1900 Irrigation Congress, attendees were split over federal involvement in reclamation projects, with Congressman Francis Newlands of Nevada heading those in favor. In 1901, federally administered irrigation found a champion in the White House. Theodore Roosevelt, an ardent conservationist, became president following the assassination of William McKinley on September 14, 1901. At that time, to be a conservationist mostly meant to be pro-reclamation. On June 17, 1902, Roosevelt signed the Reclamation Act—also known as the Newlands Act—into law.[12]

The Reclamation Act, through the sale of public lands in desert states, created "the 'reclamation fund,' to be used in the examination and survey for and the construction and maintenance of irrigation works for the storage, diversion, and development of waters for the reclamation of arid and semiarid lands in the said States and Territories."[13] Its significance was enormous. Historian

---

10. Follansbee, *A History*, 1:53.
11. Worster, *Rivers of Empire*, 156–59.
12. Rabbitt, *A Brief History*; Follansbee, *A History*, vol. 1; Worster, *Rivers of Empire*, 159–69.
13. An act appropriating the receipts from the sale and disposal of public lands in certain States and Territories to the construction of irrigation works for the reclamation of arid lands, 57th Cong, 1st sess., *U.S. Statutes at Large* (1902) 32:388–90.

Donald Worster called it "the most important single piece of legislation in the history of the West."[14] The Secretary of the Interior administered the Act, and he delegated the responsibility to USGS. It seemed a reasonable choice; by drawing upon existing expertise and infrastructure, work could begin more quickly. The newly created Reclamation Service was placed in the Division of Hydrography. With larger duties and an expanded budget, the division became a branch in its own right.[15]

Walcott asked Newell to lead the new branch and to serve as chief engineer of the Reclamation Service, with Arthur Davis as assistant chief engineer. As their guiding light, the men had the principle of conservation, albeit defined in the sense of hydrographic engineering. Watercourses needed to be manipulated to allow for absolute, complete consumption of every drop of water before it reached the ocean.

The Birth of the U.S. Reclamation Service

In January 1907, Charles Walcott left the Geological Survey to become secretary of the Smithsonian Institution. In March, the Reclamation Service left USGS, with Walcott recommending to Secretary of the Interior James R. Garfield that "the principal need is that of establishing a somewhat more direct personal contact between the Director of the Reclamation Service and the Secretary of the Interior and the cutting out of intermediate steps which are believed, through the experience of several years, to be unnecessary and to serve rather to delay than expedite public business."[16] We can assume that, given the rancor Congress heaped on Powell in the late nineteenth century, Walcott may have wanted to cut USGS out of the political discourse on Western water and land development and focus more on science.

Newell became the first director of the newly independent U.S. Reclamation Service (Reclamation), and Davis stepped into

---

14. Worster, *Rivers of Empire*, 130. For a comprehensive look at the Bureau of Reclamation, see also Donald J. Pisani, *Water and American Government: The Reclamation Bureau, National Water Policy, and the West, 1902–1935* (Berkeley: University of California Press, 2002).
15. Follansbee, *A History*, 1:83–85; Rabbitt, *Minerals, Lands, and Geology*, 2:351. For a general history of the Bureau of Reclamation, see *Brief History of the Bureau of Reclamation*, http://www.usbr.gov/history/briefhis.pdf (accessed February 21, 2006); and Michael C. Robinson, *Water for the West: the Bureau of Reclamation, 1902–1977* (Chicago: Public Works Historical Society, ca. 1979).
16. Institute for Government Research, *The U.S. Reclamation Service*, 24.

its chief engineer's position.[17] Secretary Garfield recommended 36-year-old George Otis Smith as the fourth director of USGS.[18] By the end of the First World War, water science was high on Smith's list of priorities, with the management and utilization of the Colorado River occupying a position of particular attention. This was justly deserved because a combination of national and international events continued to push water management of the Colorado River into the national spotlight. Such developments constituted sweet music to Davis, who in 1914 succeeded Newell as director of Reclamation. Davis had dreamed of controlling the Colorado River for years, and now he was in the position to make it happen.

After passage of the Reclamation Act in 1902, Davis had begun a series of public pronouncements of his grand vision of waterworks along the Colorado River.[19] By the late 1910s, the Reclamation Service was a struggling agency with a meager record, a tenuous position in the best of times but particularly so during World War I. Davis saw impressive dams and reservoirs as the bureau's salvation.[20] Politically savvy, he set out on a water-development course that showcased massive projects that would impress the nation as to what the fledgling bureau could do to solve the water problems of the arid West.

Reclamation decided that a comprehensive plan of flood control and water supply was required for the lower Colorado River and focused their energies in this region.[21] The fledgling bureau had few scientists but many engineers, and although they started to develop an overall plan for flood control and flow regulation in the Colorado River drainage, they still needed the data and expertise offered by the USGS, their former mother agency. For example, in 1914, Reclamation floated a drill rig down the Green River to below the Confluence with the Grand (now Colorado) River to investigate the footings of a potential dam site.[22] Among the incidental

---

17. George Wharton James, *Reclaiming the Arid West: The Story of the United States Reclamation Service* (New York: Dodd, Mead, 1917).
18. Rabbitt, *Minerals, Lands, and Geology*, 3:56–57. For a summary of Smith's life, see Clifford Nelson, Smith, George Otis (American National Biography Online, 2000), http://www.anb.org/articles/13/13-01537.html (accessed December 8, 2004).
19. Hundley, "The Politics of Reclamation," 297, 317.
20. deBuys and Myers, *Salt Dreams*, 159–60.
21. Stevens, *Hoover Dam*, 16–17.
22. Robert H. Webb, Jayne Belnap, and John S. Weisheit, *Cataract Canyon: A Human and Environmental History of the Rivers in Canyonlands* (Salt Lake City: University of Utah Press, 2004).

participants was a USGS hydraulic engineer named Eugene C. La Rue, who began his career when Reclamation was still a branch of the Survey. La Rue was doing his own work on hydrology of the Green and Colorado rivers for an upcoming publication on water supply.[23] La Rue would become one of the central figures in the controversies over how to regulate the Colorado River through the following decade.

The potential for hydropower production convinced public utilities to seek scientific data as well. Hydrologists and hydraulic engineers working for USGS became involved, working for private, state, and federal agencies. Some of the reports produced by USGS during this period provide the classic documentation on this river and its potential water development.[24] But tensions between the two sister bureaus developed, forming the basis of the sometimes friendly, sometimes antagonistic rivalry that has persisted into the twenty-first century.

Politics, Science, and the Colorado River

Problems kept mounting along the lower Colorado River. Mexico was in the throes of a bloody revolution, dimming the prospects for international cooperation on Colorado River control in the 1910s. President William Taft became "deeply interested in the problem that has arisen regarding the control of the Colorado River,"[25] and the Secretary of the Interior called for a Board of Review in June 1911. The resulting report urged immediate appropriation of "at least" $1 million for repair and protection of Colorado River levees and emphasized the need for continued maintenance of canals and irrigation works on both sides of the border.[26]

The authors of the report clearly were aggravated by Mexico's political turmoil but may have used that as a convenient scapegoat for their frustrations with the unruly river. In addition to negotiating water management agreements with Mexico, the report suggested "the modification of the boundary line between the United States and Mexico with a view to facilitating the solution of the entire Colorado River problem." At the time, the river swung its

---

23. Eugene C. La Rue, *Colorado River and Its Utilization* (Washington, DC: U.S. Government Printing Office, U.S. Geological Survey Water-Supply Paper 395, 1916).
24. La Rue, *Colorado River and Its Utilization*; La Rue, *Water Power and Flood Control*.
25. "Seek Light on River Control," *Los Angeles Times*, May 31, 1911 (accessed December 27, 2004).
26. The Board of Review's entire 1911 report was reproduced in Cory, *The Imperial Valley and the Salton Sink*, 1445–49.

channel across a band one-to-two miles wide,[27] which made for an ever-changing border. Rather than adjust the international boundary, the border remained the center of the river, and Taft's recommendation to Congress for appropriation of an additional million dollars was rejected.[28]

When, as a result of the All-American Canal survey and strong lobbying by Imperial Irrigation District attorney, Phil Swing,[29] Congressman William Kettner introduced a 1919 bill authorizing the canal's construction, Davis pounced. He urged rejection of the bill, arguing that the canal should only be constructed as part of a more comprehensive program, one designed by the Reclamation Service that included dams and reservoirs well upstream in the watershed. Specifically, Davis wanted a "high" (tall) dam funded by the federal government. He was joined by politicians and lobbyists who objected to the Kettner bill for primarily territorial reasons. The Imperial Irrigation District had no choice but to sign on with Davis's agenda if their canal was to be built. The original bill was jettisoned, and in May 1920, Congress approved the Kinkaid Act, which directed the Secretary of the Interior to conduct a detailed study of Imperial Valley's irrigation needs as well as those of adjacent lands, and regional water storage.[30]

Reenter USGS

The year 1920 was not the best one for USGS Director George Smith. Employees were trickling back from their military assignments during World War I, only to realize that their salaries were inadequate.[31] A significant number left USGS for more lucrative positions with private industry and at universities, while others took leave in order to take temporary employment at a higher wage. In the middle of the resulting hiring frenzy, USGS was assigned increased responsibility by Congress. Smith was almost certainly thinking of the Kinkaid Act and Davis when he lamented the 1891 cancellation of Powell's irrigation survey, "it is unfortunate that this

---

27. Ibid., 1446–49.
28. Ibid., 1449.
29. See Moeller, *Phil Swing and Boulder Dam*, for more details about Swing's involvement in the negotiations for the All-American Canal.
30. Norris Hundley, "The West Against Itself: The Colorado River—An Institutional History," in *New Courses for the Colorado*, ed. Gary D. Weatherford and F. Lee Brown (Albuquerque: University of New Mexico Press, 1986), 12–13; Hundley "The Politics of Reclamation," 310–17.
31. Rabbitt, *Minerals, Lands, and Geology*, 3:208.

special investigation was discontinued, for now authoritative information of this kind would be invaluable in planning the storage of flood waters for power and irrigation in the West."[32]

Had the irrigation survey continued as Powell envisioned it, USGS might have had far more comprehensive stream gaging records, with their essential data on water flow and availability. Dam and canal sites would be identified and practically ready for construction. As it was, there was relatively little data available for making some critical decisions.[33] To make ends meet, USGS, and particularly its Water Resources Branch, sought outside funding, offering a fifty-fifty cooperative agreement of matching state funding with federal funding.[34]

Mostly because of the single-minded efforts of one individual, USGS did possess data on the Colorado River. In 1912, Eugene C. La Rue began his investigations along the Colorado River, which culminated in his first significant monograph, *Colorado River and its Utilization*, published in 1916.[35] True to USGS form of the era, La Rue included specific management recommendations for dams, reservoirs, and canals.[36] Among other points, he concluded that the Colorado River did not carry sufficient water to meet all irrigation demands, and hence it was crucial to minimize water waste. He believed that less storage capacity might be needed in the upper basin than in the lower to control floods.

La Rue continued his research through the 1910s and early 1920s, seemingly traveling everywhere along the Colorado River and its principal tributaries. He investigated numerous dam sites, many of which would later become dams. He became convinced that a high dam built to store water for irrigation and flood control should be constructed in Glen Canyon upstream from Lee's Ferry. Such a dam, impounding the river in a narrow canyon, would create a reservoir of low surface area; consequently, evaporation would be low and less water would be wasted, considering the amount of storage that would be available. By 1920, he was a highly respected authority on the Colorado River, and La Rue was decidedly angry at being left out of the Reclamation Service plans authorized under

---

32. U.S. Department of the Interior, *Forty-First Annual Report of the Director of the United States Geological Survey to the Secretary of the Interior* (Washington, DC: Government Printing Office, 1920), 9–10.
33. Robert Follansbee, *A History*, vol. 2; Langbein, "L'Affaire LaRue" (1983):41.
34. Follansbee, *A History*, 2:2–3.
35. La Rue, *Colorado River and Its Utilization*.
36. La Rue, *Colorado River and Its Utilization*; Langbein, "L'Affaire LaRue" (1983):39–40.

the Kinkaid Act. He blamed this oversight on his former boss, Davis, who clearly had politics more than science on his mind.[37]

La Rue was determined to promote his favored high dam site, which was tucked in between the walls of Navajo Sandstone four-and-a-half miles above Lee's Ferry. He favored a 780-foot-high rock-fill dam, created by "blasting in the canyon walls" and "made watertight by sluicing fine material into the dam."[38] In 1921, he personally established the streamflow gaging station at Lee's Ferry, and his sponsor, Southern California Edison, hired local men to survey the dimensions of reservoir his dam would impound.[39] He traveled the rivers, on flat water and whitewater, searching for dam sites that met his overall plan for many small reservoirs instead of a few large ones.

Within the cacophony of voices calling for flow regulation, two began to stand out. One voice came from La Rue, who called for a basin-wide plan of dams and reservoirs designed to minimize evaporation, maximize power production, and shift flood control to the lower basin. The second voice belonged to Davis, who used the U.S. Reclamation Service as his personal megaphone to push for a single high dam and huge, multi-purpose reservoir. Davis's plan lacked some of the scientific intuition and the broad, basin-wide planning of La Rue's, and La Rue's plan lacked the political sophistication, vision, and clout of Davis's plan.[40] A high-level clash of science and policy was brewing, and Davis had the bully pulpit.

The Colorado River Compact and the Search for Dam Sites

Nearly two years after passage of the Kinkaid Act of 1920, Davis submitted a report to Interior Secretary Albert Fall, who in turn transmitted it to the Senate.[41] Davis had applied his vision of the Reclamation Service as a three-way broker between the federal government, who had the money; state and local governments, who had the individual taxpaying constituents; and the irrigation districts, which needed cheap water to make the desert bloom. His vision fed the ambition of an agency, which by the mid-twentieth century would become an enormously powerful bureaucracy that literally changed the landscape of the American West with its

---

37. Hundley, *Water and the West,* 246–47.
38. La Rue, "Tentative plan for dam," 200, 213.
39. Reilly, *Lee's Ferry,* 278–82; Follansbee, *A History,* 2:105.
40. Reilly, *Lee's Ferry,* 278–79.
41. *Problems of Imperial Valley and Vicinity,* S. Doc. 142, 67th Cong., 2d sess., 1922 ; Hundley, "The Politics of Reclamation," 317.

network of dams, canals, and power lines.[42] The first step was to get all the participants onto the same page.

The Fall-Davis report, as anticipated, urged federal construction of a high dam, now specified to be at or near Boulder Canyon and the tiny town of Las Vegas, Nevada. Davis sought to rally all the various interests behind this massive dam project, the construction of which would be unprecedented in engineering history. The government was to be reimbursed by leasing rights to the hydroelectric power generated by the dam as well as the water delivered by the All American Canal, "to be reimbursed by the lands benefited." In addition, the report recommended that future Colorado River development give top priority to river and flood control, then to storage for irrigation, and last, to power generation.[43]

The idea of generating hydroelectric power was not new, but after World War I, the issue became more urgent as the nation searched for cheap, reliable power for economic development. In June 1920, Congress approved the Federal Water Power Act, which established the Federal Power Commission and gave it authority over waterpower development projects.[44] Los Angeles, with its rapidly growing population, faced a projected power shortage in a few years, not to mention a limited water supply, in spite of its acquisition of water rights in the Owens River valley on the eastern side of the Sierra Nevada.[45] The high dam in Boulder Canyon was expected to produce an enormous amount of power, at least for that period in U.S. history, as well as regulate water deliveries to Southern California.

William Mulholland,[46] the powerful head of the Los Angeles Bureau of Water Works and Supplies, and E. F. Scattergood of the Bureau of Power and Light, saw their salvation in Davis's Boulder Canyon dam. In August 1920, the city council announced that it "does strongly favor the obtaining by the city direct from the Colorado River of such quantity of electric power as, together with other

---

42. Reisner, *Cadillac Desert*.
43. *Problems of Imperial Valley*, 21; Stevens, *Hoover Dam*, 20, 26; Hundley, "The Politics of Reclamation," 317.
44. An Act to Create a Federal Power Commission; to Provide for the Improvement of Navigation; the Development of Water Power; the Use of the Public Lands in Relation Thereto, and to Repeal Section 18 of the River and Harbor Appropriation Act, approved August 8, 1917, and for Other Purposes. 66th Cong., 2nd sess., *U.S. Statutes at Large* (1920), 41:1063–77. See Pisani, *Water and American Government*, 210–13, for a discussion of the commission.
45. Reisner, *Cadillac Desert*, 59–96.
46. For a biography of Mulholland, see Catherine Mulholland, *William Mulholland and the Rise of Los Angeles* (Berkeley: University of California Press, 2000).

power resources available to the city, will be sufficient for all future needs of its inhabitants."[47] The move did not go unnoticed by private power companies, particularly Southern California Edison, which had already investigated potential dam sites in Boulder Canyon, among others, in 1902. That project was dismissed as impractical, transmission-line technology not being adequate to the task of delivering electric power over great distances at that time. But engineering advances coupled with growing demand, made the project quite feasible by the early 1920s, and the company was determined to be included in the action.[48]

California was not the only state coveting Colorado River water; it was just the most vocal and needy. The other states in the drainage basin—Colorado, Wyoming, Utah, New Mexico, Arizona, and Nevada—had reason to fear California's ambitions, as the doctrine of prior appropriation ("first in time, first in right") favored the Golden State. The other states needed to negotiate a water pact to protect their future supply. Colorado water law attorney Delph Carpenter proposed an interstate Colorado River Compact, which Congress agreed to in August 1921 but which required the acquiescence of the individual states. The seven states that contribute drainage area to the basin needed to agree to an apportionment of the waters, but they disagreed on many of the particulars. A series of heated sessions began in January 1922, with Secretary of Commerce Herbert Hoover as chairman.[49]

Eleven months later, on November 24, 1922, delegations from the seven states agreed to a large-scale division of the drainage basin, and the Colorado River Compact was signed pending ratification by the state legislatures.[50] The dividing point between the two basins was an imaginary place on the Colorado River called Compact Point, a short distance downstream from the Utah–Arizona border at the head of Grand Canyon. Compact Point separated the Upper Basin states (Colorado, Wyoming, Utah, and New Mexico) from the Lower Basin states (Arizona, Nevada, and California). On the basis of the hydrologic studies of the U.S. Reclamation Service,[51] they believed

---

47. *Problems of Imperial Valley*, 282.
48. William A. Myers, *Iron Men and Copper Wires: A Centennial History of the California Edison Company* (Glendale, CA: Trans-Anglo Books, 1983), 177–79.
49. The firestorm surrounding the forging of the Colorado River Compact has been discussed in numerous publications, among them Hundley, *Water and the West*.
50. Stevens, *Hoover Dam*, 18.
51. The figure of 17.5 million acre-feet of water yield is quoted in Reisner, *Cadillac Desert*, 124–25.

that about 17.5 million acre-feet of water were available from the Colorado River basin, and the Upper Basin states were apportioned 7.5 million acre-feet. In turn, those states were required to provide 7.5 million acre-feet of water to the Lower Basin states, averaged over decade periods to allow for the hydrologic variability. Arizona was reserved a million acre-feet from the Gila River, and Mexico's 1.5 million acre-feet share was not apportioned until 1944.[52]

This division of the waters under the Colorado River Compact has persisted to the twenty-first century, even though the long-term average water yield of the basin was overestimated. La Rue knew that the amount of available water was overestimated; he calculated that only 15.2 million acre-feet were available.[53] Neither La Rue nor the Reclamation Service hydrologists had any idea that the period for which they had data was a highly unusual wet period, unprecedented in a nearly 1,200-years tree-ring record.[54] In fact, present-day estimates of annual flow volume past Lee's Ferry indicate that only 14.2 to 14.8 million acre-feet are to be expected.[55]

Of the various uses of the river, water itself was assigned the highest priority, followed by power generation; navigation was a distant third. After the formal signing of the Compact, Arizona had second thoughts. In 1913, James Girand, an Arizona engineer, had investigated the private construction of a dam on the Colorado River at Diamond Creek, 226 river miles downstream of Lee's Ferry.[56] The project had been creeping along in the 1910s, and it enjoyed popular support in Arizona. The Federal Power Commission was reluctant to issue the necessary licenses for the Diamond Creek Dam until negotiations for the Compact were completed, to the anger of Arizona politicians. Furthermore, Arizonans and other

---

52. deBuys and Myers, *Salt Dreams*, 160–61; Sax, Abrams, and Thompson, *Legal Control of Water Resources*, 707–8.
53. La Rue's annual flow volume at Lee's Ferry, as presented in 1925 (La Rue, *Water Power and Flood Control*, 112) gives a figure of 15.2 million acre-feet (MAF) of flow at Lee's Ferry, Arizona, for 1895–1922. Others calculate the flow during a shorter period from 1905–1922 as 16.1 MAF; Robert H. Webb, Richard Hereford, and Gregory J. McCabe, "Climatic fluctuations, drought, and flow in the Colorado River," in *The State of the Colorado River Ecosystem in Grand Canyon*, ed. Steven P. Gloss, Jeffrey E. Lovich, and Theodore S. Melis, 57–68 (Flagstaff, AZ: U.S. Geological Survey Circular 1282, 2005).
54. Connie A. Woodhouse, Kenneth E. Kunkel, David R. Easterling, and Edward R. Cook, "The twentieth-century pluvial in the western United States," *Geophysical Research Letters* 32 (2005): doi:10.1029/2005GL02413.
55. Connie A. Woodhouse, Stephen T. Gray, and David M. Meko, "Updated streamflow reconstructions for the Upper Colorado River Basin," *Water Resources Research* 42 (2006): doi:10.1029/2005WR004455.
56. Hundley, *Water and the West*, 164–65.

westerners were quite concerned about having to compete directly with California for Lower Basin allotments.[57]

The Arizona legislature refused to ratify the Compact, which meant that Congress did not ratify it either. The political wrangling would take years to resolve, and armed conflict was threatened. In the mid-1930s, by order of Arizona Governor Benjamin B. Moeur, the Arizona National Guard defended the east bank of the river near Parker against any attempt, by California or the Bureau of Reclamation, to build a structure in the riverbed.[58] Despite the threat of troops, the handwriting was on the wall for Arizona. Control—specifically, federal control—of the Colorado River was inevitable. The Arizona Legislature finally signed the Compact in 1944, allowing Congress to ratify it that year.[59]

But back in 1922, the Fall-Davis report, which advocated the Boulder Canyon dam site, would seem to have excluded USGS from using its science to mold the water-development plan for the Colorado River. The Reclamation Service controlled the lion's share of Congressional funding to study water development, and the growing interagency rivalry minimized the direct USGS role. Three other factors kept USGS in the game until the bitter end, achieved when Congress allocated funding for the Boulder Canyon Project Act in 1928. The first and foremost was the independent need for power utilities to assess the potential for hydroelectric power dams in the canyons of the Colorado River and its principal tributaries, the Green and San Juan rivers. The second was Congress itself, which insisted on hearing what USGS had to say; after all, Congress was allocating monies to USGS to assess national water resources. Finally, La Rue, the very stubborn hydraulic engineer who had a different plan for water development in the Colorado River basin, demanded to be heard. Before it was all over, La Rue very definitely was heard, and it cost him dearly.

---

57. Reilly, *Lee's Ferry*, 278.
58. Reisner, *Cadillac Desert*, 267–68.
59. Thomas E. Sheridan, *Arizona: A History* (Tucson: University of Arizona Press, 1995), 341.

3

Prelude to an Expedition

Washington and Flagstaff

In 1922, at the insistence of Arthur Davis, the U.S. Reclamation Service staked its reputation on a dam site in Boulder Canyon, putting all water into one reservoir. Elsewhere in the Colorado River basin, the surveying of potential dam sites fell to USGS, with financial and logistical support provided by private power companies, notably Southern California Edison and Utah Power and Light Company. USGS teams were able to perform much of the work from land, but simply getting to some potential dam sites required full-fledged river expeditions. The crews had to be carefully chosen. Expeditions needed at least one boatman with some knowledge of the river, as well as additional boatmen willing to work hard and learn on the job, and a science crew consisting of a topographer, at least one topographic assistant, a geologist, and a hydrologist. Most trips also had their own cook.

By the early 1920s, river travel on the Colorado River and its major tributaries was relatively well known.[1] James White may or may not have ridden through the canyon on a log raft in 1867, potentially earning the honor of first to travel through the Grand Canyon by river;[2] enough ambiguity remains as to his travels that we will likely never know for sure. We do know that John Wesley Powell was the first to systematically explore the Green River and

1. Lavender, *River Runners*.
2. Eilean Adams, *Hell or High Water: James White's Disputed Passage through Grand Canyon, 1867* (Logan: Utah State University Press, 2001).

the Colorado River downstream from its confluence with the Green River in 1869 and 1871–72. In 1889, Frank C. Kendrick explored the then-Grand River from Grand Junction to the confluence with the Green River, surveying a potential railroad route. Frank Brown and Robert Stanton picked up Kendrick's survey line and carried it—at first using survey instruments, but later, by ocular estimate—through Cataract, Glen, and Grand canyons during two trips in 1889 and 1889–90.[3] This expedition lost three men, including President Brown of the railroad company.

The publicity surrounding these expeditions built the public perception that Colorado River travel was arduous and dangerous, a perception that the little-known third Grand Canyon expedition should have dispelled. Seemingly just to do it, George Flavell and Ramon Montéz managed to avoid the toils of Powell and Stanton by running most of the rapids of Cataract and Grand canyons in 1896–97, but their experiences were largely forgotten.[4] Trappers and adventurers quietly followed with little publicity. First came Nathaniel Galloway and William Richmond, who followed Flavell and Montéz in 1896 but were more interested in obtaining beaver pelts than having an adventure. An even more obscure trip, led by a would-be prospector named Elias Benjamin "Hum" Woolley, arrived at Lee's Ferry on August 23, 1903, and launched into Grand Canyon;[5] the most important thing known about this trip was that all of the participants successfully exited the canyon at its downstream end.

Galloway repeated the journey with Julius Stone in 1909[6] and became the first person to boat through Grand Canyon twice. His boat design and boating technique became the standard for Colorado River boatmen.[7] Another adventurer, Bert Loper, traveled through the treacherous Westwater and Cataract canyons,[8] but he stopped at the entry to Grand Canyon in 1908, left behind by his fellow travelers Charles Russell and Edwin Monett. Loper ended up

---

3. Webb, Belnap, and Weisheit, *Cataract Canyon;* Robert H. Webb, *Grand Canyon, A Century of Change: Rephotography of the 1889–1890 Stanton Expedition* (Tucson: University of Arizona Press, 1996).
4. George F. Flavell, *The Log of the Panthon: An Account of an 1896 River Voyage from Green River, Wyoming to Yuma, Arizona through the Grand Canyon*, ed. Neil B. Carmony and David E. Brown (Boulder, CO: Pruett Publishing, 1987).
5. Reilly, *Lee's Ferry*, 191.
6. Julius F. Stone, *Canyon Country: The Romance of a Drop of Water and a Grain of Sand* (New York: G. P. Putnam's Sons, 1932).
7. Lavender, *River Runners*, 36–38, 40–43.
8. Webb, Belnap, and Weisheit, *Cataract Canyon*, 27–28; Lavender, *River Runners*, 45.

rowing and towing his boat back upstream through most of Glen Canyon. Ellsworth and Emery Kolb, who ran a photographic studio at the South Rim of Grand Canyon, repeated the route of Powell's 1869 expedition in 1911 with a goal of taking still photographs and making movies of the river. Ellsworth wrote a book publicizing their trip,[9] and the brothers began projecting a movie about their adventures at their studio.[10]

By 1921, only twenty-seven river runners had traveled the entire length of Grand Canyon (not including James White).[11] Three others—Frank Brown, Peter Hansbrough, and Henry Richards—had drowned during the ill-fated 1889 expedition because they were not wearing life preservers.[12] Not including several general magazine articles written by Stanton and others, only three books provided detailed descriptions of river travel in Grand Canyon: Powell's account, which Stanton knew contained considerable exaggeration and inaccuracy; Frederick Dellenbaugh's *A Canyon Voyage*,[13] essentially a revision of Powell's story; and Ellsworth Kolb's book, which is relatively accurate. Public perception followed Powell's version: river travel was dangerous, even death-defying. When USGS proposed to conduct a river expedition through Grand Canyon to complete its studies of potential dam sites in 1923, a national audience became keenly interested.

The Preliminaries: River Trips in 1921 and 1922

Funded primarily by public utilities searching for hydropower opportunities, and prodded along by Eugene C. La Rue and the perceived need to control the Colorado River "menace," USGS undertook river expeditions in 1921 and 1922 to survey potential dam sites in the Upper Basin. These expeditions traveled by land to get to the few readily accessible sites, and by boat on the otherwise inaccessible reaches of the upper Colorado River and its major

---

9. Ellsworth Kolb, *Through the Grand Canyon from Wyoming to Mexico* (New York: Macmillan Company, 2nd ed. 1927). Original published in 1914; see Richard Quartaroli, appendix to Ellsworth L. Kolb and Emery C. Kolb, *The Brave Ones: The Journals and Letters of the 1911–1912 Expedition Down the Green and Colorado Rivers*, ed. William C. Suran (Flagstaff, AZ: Fretwater Press, 2003).
10. Kolb, *Through the Grand Canyon*; Emery and Ellsworth Kolb, *Grand Canyon Film Show*.
11. Richard D. Quartaroli, "Chosen Sons of the Gods: Grand Canyon River Runners List," computer printout dated May 18, 1995.
12. Smith and Crampton, *The Colorado River Survey*, 74–84.
13. Frederick C. Dellenbaugh, *A Canyon Voyage* (New Haven, CT: Yale University Press, 1908).

tributaries, the Green and San Juan rivers, all of which included some rough water with challenging rapids. These surveys would link other surveys of dam sites previously made along less-dangerous flat-water reaches; for example, Reclamation had investigated the Junction Dam site, downstream from the Confluence of the Green and Grand (Colorado) rivers, in 1914.[14]

In the early 1920s, instrumental surveys—using an alidade and stadia rod—were the only means to obtain accurate map data. These techniques were painstaking and laborious; each topographic point was obtained by observing the stadia rod from an instrument station and making calculations of distance that accounted for vertical angles. The surveying instrument had to be set up on a previously obtained survey point, leveled, and oriented, which took time and care; if the instrument fell out of adjustment or was not carefully used, considerable error would creep into the topographic map. Overall, the survey had to be tied in to existing benchmarks, which was difficult to accomplish in the vertical and labyrinthine river canyons of the Colorado Plateau. By the 1930s, the evolving science of photogrammetry and aerial photography would ultimately automate the production of topographic maps, but for the mission at hand, ground surveys were needed.

The San Juan River, 1921

Downstream from its headwaters in southwestern Colorado, most of the course of the San Juan River passes through open country in northwestern New Mexico and southeastern Utah. A few river miles downstream from the town of Bluff, Utah, the San Juan River enters the first of two canyons that ultimately wind to a confluence with the Colorado River in Glen Canyon. Legendary landscape photographer William Henry Jackson explored the entrance to this canyon in 1875, or five years before settlement of Bluff by the equally legendary Hole-in-the-Rock expedition.[15] A minor gold rush in the early 1890s encouraged boat travel by prospectors,[16] but much of this history is fragmentary. For USGS to explore the canyons of the San Juan by boat, they needed some nongovernmental expertise.

---

14. Webb, Belnap, and Weisheit, *Cataract Canyon*, 29–30.
15. Douglas Waitley, *William Henry Jackson, Framing the Frontier* (Missoula, MT: Mountain Press Publishing Company, 1999), 153–55; Robert S. McPherson, *A History of San Juan County: In the Palm of Time* (Salt Lake City: Utah State Historical Society, 1995), 90, 97–101.
16. James M. Aton and Robert S. McPherson, *River Flowing from the Sunrise: An Environmental History of the Lower San Juan* (Logan: Utah State University Press, 2000), 113–20.

USGS turned to a middle-aged Bert Loper to serve as head boatman. Loper had nearly three decades of river experience by the time the trip launched, including the first descent through Westwater Canyon on the Colorado River.[17] Kelly Trimble, a USGS topographical engineer, headed the expedition, which was funded by Southern California Edison. Both Trimble and geologist Hugh Miser kept diaries of the expedition. The crew also included Loper's friend Elwyn Blake, a novice river runner. Blake had hired on as a rodman, but on the first day of the expedition, he took to the oars when one of the boatmen lost his nerve.[18] The men launched a few miles downstream of Bluff in July 1921 and surveyed for dam sites in 133 miles of its canyons, to its confluence with the Colorado River. The trip was productive and relatively uneventful; the participants appeared to get along and work together on the nearly three-month journey. After the Trimble expedition, river running on the San Juan River would not become a regular occurrence until the pioneering commercial river runner Norman Nevills made his first run in 1934.[19]

Cataract Canyon, 1921

In late June, another USGS expedition, also funded by Southern California Edison, launched from Green River, Utah, into Labyrinth and Stillwater canyons on the Green River. This expedition, headed by topographer William Chenoweth, was to pick up the 1914 survey at the Confluence and carry it through Cataract and Glen canyons to Lee's Ferry. Chenoweth's crew included veteran boatman Ellsworth Kolb, as well as Leigh Lint, a 19-year-old rodman who also did some rowing on the trip. Lint had worked with Chenoweth the previous year during similar dam-site surveys on the Snake River in Hell's Canyon. Geologist Sidney Paige and hydraulic engineer La Rue rounded out the ten-man crew. Emery Kolb and an assistant accompanied the trip in order to add to the Kolbs' stock of river photography.

On September 15, the men rowed into Cataract Canyon, surveying its water-surface profile and numbering its rapids.[20] Traveling on relatively low water, their expedition was fairly uneventful until

---

17. Ibid., 58–59.
18. Blake replaced Hebe Christiansen as boatman. Westwood, *Rough-Water Man*, 3–61.
19. Nevills became one of the first to run commercial river trips in the Colorado River basin. See Nancy Nelson, *Any Time, Any Place, Any River: The Nevills of Mexican Hat* (Flagstaff, AZ: Red Lake Books, 1991), 3–5.
20. Webb, Belnap, and Weisheit, *Cataract Canyon*, 29–30.

they reached Dark Canyon, near the downstream end of Cataract Canyon. La Rue documented the Dark Canyon dam site, and Ellsworth Kolb pinned one of their boats against a rock and flipped another in Dark Canyon Rapid.[21] After taking two days to pull the pinned boat out, the expedition completed its journey. Exiting from the short Narrow Canyon into Glen Canyon, and completing the survey at the mouth of the Dirty Devil River, members of the Chenoweth expedition became the 63rd through 70th river runners to pass through Cataract Canyon.[22]

At the confluence of the San Juan and Colorado rivers, the Chenoweth and Trimble expeditions met, sent some of their participants out, and combined the remainder for a side-canyon survey down to Lee's Ferry. In his diary, La Rue noted, "from the appearance of the men in the Trimble Party, they must have had a strenuous time on the San Juan River Survey. They had about one day's rations left."[23] Two brief but notable encounters took place: Blake and Lint met for the first time, and this may also have been the first meeting between La Rue and Loper.[24]

Green River Canyons, 1922

In 1922, Trimble led another river expedition, this time down the Green River from Green River, Wyoming, to Green River, Utah.[25] Ralph Woolley served as hydraulic engineer and John Reeside as the geologist, and the boatmen included Bert Loper, Elwyn Blake, and Leigh Lint. The trip, funded by Utah Power and Light Company, sought to investigate previously selected dam sites as well as identify potentially new ones. The Green River had been partially surveyed in 1914 and 1918. The Trimble expedition's goal was to link this previous work together.

Launching July 15, 1922, the men took 62 days to travel 387 river miles, passing through numerous canyons, including Canyon of Lodore, and Desolation and Gray canyons. The crew stranded boats on rocks, learned to repair the holes in their wooden boats, and

---

21. Westwood, *Rough-Water Man*, 57–59.
22. Webb, Belnap, and Weisheit, *Cataract Canyon*, 35–36.
23. Eugene La Rue, Green River, Utah, to Lee's Ferry, Arizona river trip diary, entry for October 5, 1921, box 114, folder 2, La Rue Collection.
24. Unsubstantiated rumors hold that La Rue and Loper had a serious argument when they met here, which later affected Birdseye's choice of a head boatman for the 1923 Grand Canyon expedition. Loper biographer Brad Dimock has been unable to find any contemporary evidence of this alleged dispute.
25. Westwood, *Rough-Water Man*, 73–119.

learned how to "nose" the boats along rapids—wading along shore while holding the nose of the boat. It was excellent preparation for the three boatmen to tackle a harder reach of whitewater, but only two would make the run through Grand Canyon the following year.

Glen Canyon, Western Grand Canyon, and Boulder Canyon, 1922

In 1922, La Rue, working for USGS, saw another opportunity to push his agenda for the Lee's Ferry high dam. He was now chairman of the Arizona Engineering Commission, which was charged with determining the potential for irrigation-diversions of Colorado River water.[26] He arranged a special float trip in Glen Canyon that would function both to make detailed surveys of potential dam sites and build advocacy for his dam.[27] Participants included a host of dignitaries and engineers: Clarence Stetson, secretary of the Colorado River Commission and assistant to Secretary of Commerce Herbert Hoover; John Widtsoe, a former University of Utah president and Latter-day Saints apostle; Arthur Davis; Claude Birdseye, chief engineer of USGS topographic division; Herman Stabler, chief engineer of USGS Land Classification Board; and a handful of other engineers. Lewis Freeman, a boatman, adventurer, and author who was keenly interested in reporting on the Colorado River water situation, also participated. Six crewmen, including Freeman, launched the boats at Lee's Ferry and slowly moved upriver to Hall's Crossing, where they met the rest of the party on September 8. From there, the group proceeded back downriver, enjoying the scenery and reviewing the most promising dam sites.

When the men arrived at La Rue's proposed dam site, 4.6 miles upstream from Lee's Ferry, the hydraulic engineer pulled out a huge map. As Arthur Davis and the other dignitaries huddled around, "a chorus of ribald yells"[28] echoing down the canyon suddenly interrupted La Rue's hard-earned moment in the spotlight. Several of the members of the party, including a civil engineering professor from the California Institute of Technology, had lagged behind. They appeared dressed as pirates, replete with biscuit-tin earrings and a Jolly Roger flying from the bow of their boat. After their performance of "Fifteen Men on a Dead Man's Chest," none other than Arthur Davis stepped forth and belted out the lyrics

---

26. Rabbitt, *Minerals, Lands, and Geology*, 3:233.
27. Reilly, *Lee's Ferry*, 293–95.
28. Freeman, *Down the Grand Canyon*, 209.

from "Oh, Better Far to Live and Die": "I am a Pirate King! And it is, it is a glorious thing, to be a Pirate King!" Had the hapless La Rue been able to counter with "I Am The Very Model Of A Modern Major General," with "its many cheerful facts about the square of the hypotenuse," he might have been able to rescue the moment, but La Rue's repertoire apparently did not include Gilbert and Sullivan's *The Pirates of Penzance*. The rest of his lecture fell on distracted ears.[29]

Following their return to Flagstaff, some of the men, including La Rue, Davis, Stabler, and Freeman, continued onward by road to investigate James Girand's proposed dam site at Diamond Creek. Girand and W. S. Norviel, Arizona's Colorado River commissioner, served as guides for that part of the tour. Arthur Davis had his turn a few days later. By this time in 1922, the Reclamation Service had shifted its primary dam site from Boulder Canyon downstream to Black Canyon because of the latter's superior topography and geology. Davis proudly pointed to the hard rhyolite walls where he proposed a massive masonry dam. There were no pirate attacks, and Davis appeared to have won another round with La Rue.[30]

The Colorado River Compact had been signed, and Davis still firmly controlled the conversation about how and where to regulate the Colorado River. But in March 1923, the nation acquired a new secretary of interior, physician Hubert Work. In response to the many fiscal complaints leveled at the Reclamation Service, Work hired D. W. Davis (no relation to Arthur Davis) as an adviser. D. W. Davis was president of the Western States Reclamation Association, the former governor of Idaho, and a banker—in other words, he was a bureaucrat and a politician, but not a scientist.

In June, Work summarily fired Arthur Davis, replaced him with D. W. Davis, and changed the name of his agency to the Bureau of Reclamation. Arthur Davis may have been politically astute, but even he could not weather the new secretary of the interior or the political currents circulating the West, where the Colorado River basin—Arthur Davis's myopic focus—was only one of many basins in need of federal dollars for flow regulation. Among Davis's claims were that his federal government plans for tall, power-generating dams on the Colorado River were at odds with similar dams proposed by

---

29. Freeman, *Down the Grand Canyon*, 208–10; A. R. Mortensen, ed., "A Journal of John A. Widtsoe: Colorado River Party, September 3–19, 1922, preliminary to the Santa Fe Conference which framed the Colorado River Compact," *Utah Historical Quarterly* 23 (July 1955): 195–231.
30. Freeman, *Down the Grand Canyon*, 210–15; Reilly, *Lee's Ferry*, 294–95.

private firms; the concerns of those firms and their financial backers led in part to his dismissal.[31] The removal of Arthur Powell Davis generated a firestorm in engineering circles, which bristled at the notion that engineers could not be administrators. While Work claimed he was just trying to rescue Reclamation, Davis cried foul. He was down, but he was not out; in the coming years, his influence would continue.

USGS and others had identified potential dam sites in Utah, Colorado, and New Mexico (some of which would be later used in the Colorado River Storage Project in the 1950s). USGS, and especially La Rue, was not yet finished, and Southern California Edison, its principal financial backer, still wanted to know the potential dam sites on the Colorado River in Arizona. That left the most difficult reach, and the one with potentially the most dam sites: Grand Canyon.

The Preparations for Grand Canyon, 1923

Grand Canyon has always inspired a combination of wonder and terror. The first Europeans to see it, under the leadership of Spanish explorer García López de Cárdenas in 1540, found it baffling in its perspective, initially estimating the width of the river within the depths of the canyon to be a paltry six feet.[32] With his eloquent prose, Powell broke the spell of terra incognita but added to the perception that the canyon was either impassable or that one had to be extremely lucky to survive the trip. Others had proved the trip was not as dangerous as Powell claimed, but they had not received nearly the publicity. If, for example, Flavell had not died prematurely and shortly after his 1896 joyride through the canyon, the public might have gained the impression that a Grand Canyon trip was no more dangerous than a trip to an amusement park. Instead, to the general public, the canyon evoked images of toil, danger, and death, despite the relatively few fatalities that had actually occurred.[33]

---

31. Donald J. Pisani, *Water and American Government: The Reclamation Bureau, National Water Policy, and the West, 1902–1935* (Berkeley: University of California Press, 2002), 137–39; Charles A. Bissell and F. E. Weymouth, "Memoirs of Deceased Members: Arthur Powell Davis, Past-President, Am. Soc. C.E.," *Transactions of the American Society of Civil Engineers* 100 (1935): 1582–91; Gene M. Gressley, "Arthur Powell Davis, Reclamation, and the West," *Agricultural History* 42, no. 3 (1968): 241–57.
32. C. Gregory Crampton, *Land of Living Rock, The Grand Canyon and the High Plateaus: Arizona, Utah, Nevada* (Layton, UT: Peregrine Smith, 1972), 68–69.
33. Webb, Belnap, and Weisheit, *Cataract Canyon*, 34–37; T. M. Myers, C. C. Becker, and

The USGS expedition through Grand Canyon would accomplish two major goals. By comprehensively surveying the river corridor, USGS would locate the last potential dam sites in the Colorado River system, including any that might compete with the Boulder (now Black)[34] Canyon site promoted by Davis and Reclamation. And because it would thoroughly map the canyon, the expedition would essentially eliminate the river's mystique, reducing John Wesley Powell's "great unknown" to a series of maps and profiles. Almost by accident, it would be the last of the great exploratory expeditions in the American West, marking the turn from the frontier image of the nineteenth century to modernity and a sense of known topography on the Colorado Plateau. Subsequent Grand Canyon river trips would rely on novelties or gimmicks—a staged disappearance, a motion picture project, the participation of women, a dog, and a bear cub—to get even a small part of the publicity generated by the 1923 USGS expedition.[35]

In 1923, a Grand Canyon river expedition was news, and exciting news at that. Other Grand Canyon news deemed worthy of reporting in that year included whether or not the postmaster should be permitted to keep his dog within the newly created national park, a matter that went all the way to President Warren Harding's desk.[36] USGS administrators and scientists, particularly La Rue, wanted to exploit the publicity to get their plans for the Colorado River development into the public eye. No one—not USGS, not the Department of Interior, the Reclamation Service, or the general public—paid any attention to the environmental effects that dam construction would have on Grand Canyon National Park, much less on the river as a whole. The publicity focus was entirely on the adventure, and risk, of river running, combined with the science of designing flow regulation and producing hydropower from the Colorado River.

With all this in mind, USGS released a carefully worded "Memorandum for the Press" in which they noted: "A series of special

---

L. E. Stevens, *Fateful Journey, Injury and Death on Colorado River Trips in Grand Canyon* (Flagstaff, AZ: Red Lake Books, 1999).

34. The first site proposed for the Boulder Canyon dam was indeed in Boulder Canyon. When testing—conducted in 1921—revealed the site was geologically unsuitable, testing began downstream in Black Canyon. Although the dam was ultimately built in Black, not Boulder, Canyon, the site was still frequently referred to as Boulder Canyon.
35. Lavender, *River Runners*, 66–84.
36. "Dog in Grand Canyon Debated by Cabinet," *New York Times*, January 13, 1923 (accessed February 27, 2006).

stream surveys, showing plans and profiles of streams and sufficient topography to cover all possible locations for structures needed in a comprehensive scheme of water development, was begun in the Colorado River Basin in 1909. About 1,200 miles of the Colorado and Green rivers and several hundred miles of their principal tributaries had been mapped by the end of 1922 ... [On the 1923 Grand Canyon survey] detailed examinations will be made, of possible dam sites, which will be considered from both the engineer's and the geologist's point of view, and the surveys will be of value not only in connection with the possible building of dams, but in expediting classification of lands and contributing to the hydrographic and geologic knowledge of the country and will be available for incorporation in future maps and reports on the region."[37]

Expedition Personnel

Trip Leader, Claude Birdseye

USGS selected Claude Hale Birdseye, its chief topographic engineer, to organize and lead the survey. The forty-five-year-old Birdseye, the oldest member of the expedition, had an accomplished field record and was regarded as a brilliant topographic engineer. Born in Syracuse, New York, he was a distant cousin of Clarence Birdseye of frozen-foods fame; his father was a postal inspector. An athletic man, he played football in college.[38] Upon his graduation from Oberlin College in 1901, Birdseye joined USGS as a field assistant while simultaneously pursuing graduate studies at the University of Cincinnati. Three years later, he married Grace Whitney, and ultimately they had three children. Birdseye's promotions were rapid, as his supervisors noted his "rare judgment in adopting methods and in circumventing obstacles that would have puzzled more experienced topographers."[39]

Early in his career, Birdseye undertook the challenging assignment of mapping Hawaii, including the active Kilauea volcano, from 1909 to 1912. In 1913, he climbed Mt. Rainier in a blizzard while leading a topographic survey. During World War I, Birdseye began his service as a major with the Corps of Engineers and ended

---

37. "Department of the Interior, Memorandum for the Press: Surveys of Grand Canyon of the Colorado," n.d., NARA, Record Group 57, Records of the Topographic Division.
38. Howard H. Martin, "Conquers Grand Canyon: Bro. Birdseye Charts the Gorge," *Cadeceus of Kappa Sigma*, date/volume unknown, 322–27. A staple-bound copy of this article is held at NARA, Record Group 57, Records of the Topographic Division.
39. Untitled promotion letter, June 11, 1908, Birdseye Personnel File.

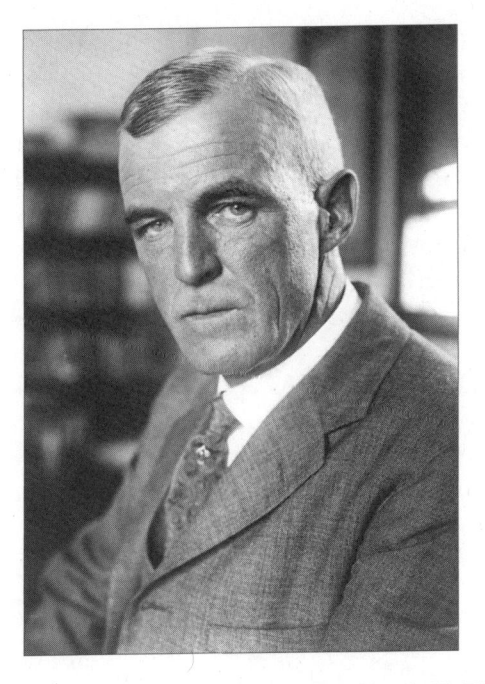

Expedition leader and topographic engineer Claude H. Birdseye, in a studio portrait taken on November 2, 1923.

Underwood & Underwood photograph, 34970RU, courtesy of the Library of Congress.

---

as a lieutenant colonel with the Coast Artillery in France. Upon his return to USGS in 1919, he was promoted to chief topographic engineer, in charge of the entire Topographic Division. The six-foot-tall, affable Birdseye was popular with his colleagues, who rewarded him with great loyalty. An astute civil servant, he had the full support of his superiors.[40]

Birdseye wanted to keep the size of the crew as small as possible and undoubtedly left behind several disappointed would-be river runners. In a post-trip radio address, Birdseye quoted a letter he received from "an old prospector in Colorado": "Dear Sir, I have noticed in Denver papers that you are organizing a party to shoot the rapids of the Colorado River. I have had considerable experience in the use of dynamite and think I would be a valuable member of your party. Please let me know if you can make use of

---

40. See the Birdseye Personnel File for Claude Birdseye's many promotions and positive comments; Ronald M. Wilson, "Claude Hale Birdseye, M. Am. Soc., C. E.," *Transactions of the American Society of Civil Engineers* 106 (1941): 1549–53.

my services." Birdseye noted, "Of course the offer was not accepted, but there were times when we wished this old prospector had preceded us and blown some of the rocks from the channel."[41]

Birdseye carefully selected the trip participants. Some of the choices were obvious—La Rue would not be left behind—but some of the choices were politically charged. Other appointments, such as rodman and junior boatmen, also were relatively easy. Birdseye engaged his cousin, Roger Birdseye, to handle the logistics of the resupplies that would be needed for the long expedition. Perhaps the most difficult decision Birdseye had to make to fill out his personnel list was the choice of head boatman. One way or the other, he would have to make compromises, and it is unknown why some individuals were selected over others. Birdseye's choices created an expedition that knew considerable disscusion, even if it did succeed in its overall goals.

Head Boatman, Emery Kolb

Why Birdseye settled on Emery Kolb to be his head boatman is a mystery we may never resolve. Ellsworth Kolb was the head boatman for the 1921 Cataract Canyon survey, and Bert Loper had been the head boatman on the 1921 and 1922 expeditions on the San Juan and Green rivers, respectively. Emery had merely tagged along in Cataract Canyon to secure photographs and movies, and he had no prior employment with USGS. Ellsworth was the sole author of the book on the brothers' 1911–12 exploits,[42] which might have given him a potential edge, but he also had a significant boating accident in Dark Canyon Rapid at the end of the Cataract Canyon expedition. Emery led the extraction of the pinned boat, which may have impressed Chenoweth and influenced Birdseye's choice. Alternatively, Emery might have been chosen simply because of agreements between the brothers; Ellsworth got the Cataract job and now it was Emery's turn, and one of the brothers had to stay at South Rim to run the photographic studio during the busy summer season of 1923.

Loper justifiably thought he had earned the head boatman job for the 1923 expedition in Grand Canyon. He felt somewhat paternal about Blake and Lint, later noting, "you might say—that they were children of mine—from a boating standpoint."[43] But Loper

---

41. Birdseye, "Radio in the Grand Canyon."
42. Kolb, *Through the Grand Canyon*. Ellsworth used some of Emery's journal in the book; Kolb and Kolb, *The Brave Ones*.
43. Otis Marston, undated notes typed from Bert Loper letter dated December 19, 1947, box 280, folder 35, Marston Collection.

Head boatman Emery C. Kolb at Diamond Creek in early October, 1923.
Emery Kolb Collection, NAU.PH.568-5158, courtesy of the Cline Library, Northern Arizona University.

had two big problems: he was fifty-three years old, and he had never boated through Grand Canyon. Birdseye clearly had Emery Kolb as his number one candidate and may have written Loper off as being too old, which Loper later claimed came from La Rue.[44] Birdseye may also have opted to offer the position first to Emery Kolb on the basis of Kolb's previous Grand Canyon experience,

---

44. Regarding his not being selected as a boatman on the 1923 trip, Loper lamented, "Mr. E. C. La Rue ... went to the Colonel (Birdseye) and told him that I was too old to handle the job, so I was left out;" Bert Loper, "Thoughts That Come to Me in the Still of the Night," undated manuscript, box 121, folder 29, Marston Collection. In another account, he noted: "If I had ever been called on and failed it would have been different but being the oldest Boatman and be left out hurt a little but it is all over now so I must turn my attention to something else for this summer." Bert Loper to H. D. Miser, March 26, 1923, box 19, folder 33, Marston Collection.

with Loper as second choice should Kolb decline.[45] Whatever the case, Loper was not given any slot on the trip, head boatman or otherwise, suggesting that at least some political influence bore on this decision.

Emery Clifford Kolb was born in 1881 in Smithfield, Pennsylvania. His was an active childhood, with adventures sandwiched between the prayers and Bible study mandated by his devoutly Methodist parents. Ellsworth, Emery's older brother, headed west, eventually settling at the South Rim of Grand Canyon, where Emery joined him in 1902. They purchased a photographic business and set up shop on the rim, eventually building a studio that clung to the precipice next to the Bright Angel trailhead. From their perch, they photographed mule parties as they prepared to head down the trail. Emery—short, lithe, and energetic—would run the four and one-half miles down the trail to the nearest clear water source in order to have prints ready to sell to the dusty, sore, souvenir-inclined tourists before they returned to the rim.[46]

The trail photographs alone did not generate enough money to pay the bills, and the Kolbs had to stay competitive. Grand Canyon attracted professional photographers, many of whom had better technical and artistic skills than the Kolbs. The brothers realized that while many photographers could capture scenic vistas, they had both the daring and the skill to obtain breathtaking action shots. Their 1911–12 river trip[47] proved that. Although they made the trip as much for the adventure as anything, "the success of the expedition," wrote Ellsworth, "depended on our success as photographers."[48] The Kolbs made a point to photograph many of the same views made famous by the Powell expedition, adding a copyright symbol to the photograph. This enabled them to cash in on familiarity while inadvertently establishing themselves as the first people to deliberately replicate photographs within the western United States.[49] Their motion pictures were the first made of a Grand Canyon river journey, and

---

45. Hugh Miser to Bert Loper, March 13, 1923, box 19, folder 33, Marston Collection.
46. Sarah Pedersen and the editors of Discovery, with the assistance of Joan Lutes, Bob Sorgenfrei, and Rick Startzman, *Emery Kolb: A Guide to the Kolb Collection in the NAU Libraries* (Flagstaff, AZ: Northern Arizona University Libraries, 1980), 1–4; Suran, "With the Wings of an Angel" (accessed September 29, 2004).
47. The diaries and letters associated with the Kolbs' 1911–12 river trip have been collected into a volume edited by Suran, Kolb and Kolb, *The Brave Ones*.
48. Kolb, *Through the Grand Canyon*, 3.
49. Webb, Belnap, and Weisheit, *Cataract Canyon*, 28–29.

they were justifiably proud of that.[50] Ellsworth's book sold tens of thousands of copies.[51]

Emery and Ellsworth, while both colorful and charismatic, were fundamentally different people. Ellsworth was relatively relaxed, a risk taker, and a ladies' man; Emery was a devoted family man, more cautious than his brother, and an emotional ball of fury. By 1915, the brothers' relationship was fragile at best, and they took turns managing the studio. Both were eager to participate in USGS expeditions to augment their river movies. In 1922, the tensions reached a head, and the brothers ended the business partnership, though Ellsworth still helped out when necessary.[52]

The early 1920s were a stressful time for Emery. In 1919, Congress changed Grand Canyon from a national monument into a national park. The change in status brought new regulations and fees, which were understandably unwelcome to residents and extant businesses. For a small family firm like the Kolb Brothers Studio, the National Park Service was just another large, well-funded organization threatening its livelihood, much like the Fred Harvey Company–Santa Fe Railroad partnership that was the primary park concessionaire. When Birdseye contacted him about the 1923 USGS trip, Kolb wanted the job but was in no mood to become subservient to another government agency.

To Birdseye's telegram asking if he would serve as head boatman and to state his salary, Kolb responded with a desired salary of one thousand dollars, and "exclusive film privilige [sic]," adding he was certain that with his ability, he could reduce both the length of the trip and the number of men needed.[53] Birdseye rejected Kolb's offer, specifically the portion granting him exclusive motion-picture rights, and noted that if he would not place the success of the journey over his own interests, he would find someone else. Kolb, backed into a corner, responded, "Handling boats excluding motion camera approved provided salary sufficeint [sic] for risk and worthy of my experience please quote."[54] Birdseye's final telegram offered a salary of five hundred dollars per month; the use of

---

50. Kolb and Kolb, *The Brave Ones*.
51. According to historian Richard Quartaroli, sales of Kolb's book probably numbered between 30,000 to 50,000 at that time. Richard Quartaroli, "Over the Edge—Over 100K in Print!," *Boatman's Quarterly Review* 19 (Spring 2006): 23.
52. Suran, "With the Wings of an Angel" (accessed October 6, 2004).
53. Emery Kolb to Claude Birdseye, February 9, 1923, box 5, folder 653, Kolb Collection, Cline Library, Northern Arizona University, Flagstaff.
54. Emery Kolb to Claude Birdseye, n.d, box 5, folder 655, Kolb Collection.

Boatman and author Lewis R. Freeman bailing out the *Grand* at the foot of Waltenberg Rapid, September 5, 1923.

Emery Kolb Collection, NAU.PH.568-5118, courtesy of the Cline Library, Northern Arizona University.

cameras was not mentioned. Kolb accepted Birdseye's offer, but the negotiations were actually just beginning.[55]

Emery had married Blanche Bender in 1905, and Blanche was well-adapted to Emery's life of adventure travel and landscape photography. Edith, their sole child, was an adventurous, newly-minted sixteen-year-old in July 1923. Raised at the Kolb Studio, she was at ease in the outdoors and in front of a camera. Both wife and daughter accompanied Emery on at least some of his canyon adventures; as a toddler, Edith was placed in a box lashed to the side of a mule on one trip, and rode a mule as an older child.[56] Edith was all too happy to join her father at Lee's Ferry, Hance Rapid, and Hermit Rapid during the 1923 expedition.

Boatman and Publicist, Lewis Freeman

Lewis Ransome Freeman was an odd choice for boatman, particularly given the availability of Bert Loper. Freeman was overweight,

---

55. Claude Birdseye to Emery Kolb, February 14, 1923, box 5, folder 656, Kolb Collection.

56. Suran, "With the Wings of an Angel" (accessed September 29, 2004).

forty-four years old, and had little whitewater experience. He may have been invited on the 1923 expedition at the suggestion of La Rue, in large part because Freeman filled two important niches. He had a demonstrated ability to row a boat on flatwater, and, more importantly, he was an expert at publicity, particularly if he took center stage. Freeman was born in 1878 in Genoa Junction, Wisconsin; his family moved to California soon thereafter.[57] His father had been a 49er, made a fortune selling lumber in Wisconsin, and then worked in manufacturing and real estate in Pasadena. When his father died in 1902, his estate totaled fifty thousand dollars,[58] which is the equivalent of over $1 million by today's standards.

Freeman graduated from Stanford University, where he was a champion athlete, lettering in football, track, baseball, and tennis. He played tennis competitively after college, and, at the beginning of the century, was considered the best player in Southern California. Freeman perfected writing in the breathy, polysyllabic style favored by many at the time, and he dabbled in journalism, largely as a foreign correspondent. In 1905, he landed an assignment as a Russo-Japanese war correspondent. He became a proficient photographer, using his still images to illustrate his magazine articles and motion pictures on his successful lecture tours.

Freeman covered World War I as a war correspondent, then launched a career as an adventure-travel writer. Many of his escapades involved boating and rivers. By 1923, he had lost his athletic build, but he had gained a wide reputation as a popular author whose articles appeared in major magazines and whose books were offered by a New York publisher. He was eight months younger than Birdseye, making him the second oldest member of the expedition. Freeman was extremely well-connected; he counted as friends Leonard Huxley, General John Pershing, and Rudyard Kipling.[59] Birdseye, Stabler, and La Rue had all been in his company on the 1922 Glen Canyon trip. He spent the months preceding the expedition finalizing the manuscript for his forthcoming book, *The Colorado River: Yesterday, To-day and To-morrow*, which

---

57. For a biography of Freeman, see Axel Friman, "Lewis Ransome Freeman—a Swedish American Author," *Swedish American Genealogist* 1 (1981): 141–43.
58. "Court Notes: Good Estate," *Los Angeles Times*, September 26, 1902 (accessed May 27, 2005).
59. Freeman spent years preparing an illustrated, limited edition version of a Kipling poem; Rudyard Kipling and Lewis R. Freeman, *The Feet of the Young Men* (Garden City, NY: Doubleday, Page, 1920).

Boatman Leigh B. Lint strikes a pose, probably on a 1922 river trip. This pose is reminiscent of William S. Hart, a cowboy actor who appeared in dozens of silent movies from 1907 through 1925.

Courtesy of George Lint.

included a discussion of the Davis-La Rue feud.[60] Freeman was to prove that including an overweight boatman on a Grand Canyon expedition was worth the burden placed on the rest of the crew. If La Rue thought that Freeman would blindly promote his plan for Colorado River development, La Rue was mistaken.

Boatman, Leigh Lint

The other two boatmen were easy choices, as they had already worked on previous USGS expeditions. Twenty-year-old Leigh Brinton Lint likely came to mind immediately. Lint was named for his birthplace of Leigh, a whistle-stop in Nebraska. His family moved every few years, landing in Weiser, Idaho, an agricultural community on the Snake River, northwest of Boise, around 1910. Lint had just graduated from Weiser High School when he hired on as a recorder, rodman, and boatman with Chenoweth's 1920 Snake River survey. He worked the following two summers

---

60. Freeman, *The Colorado River*, 417–37.

for USGS, in Cataract Canyon and on the Green River. He enjoyed river running so much that in December 1922, he and a friend launched a pleasure trip down the Snake River from Weiser to Lewiston in a boat Lint built following the design of those used on his government trips.[61]

River survey work suited Lint, who far preferred the outdoors to an office. A newspaper article reported, "Lint, though but a youth, is one of the most experienced and trustworthy boatmen for treacherous river travel in the United States. He is almost a six-footer, weighs 190 pounds, is powerfully built, and has the courage of a lion."[62] Lint stood with the confidence of a man aware of his strength. Indeed, on an employment form, he listed "physique" as one of his "special qualifications,"[63] clearly setting himself apart from the likes of Freeman. His employers and fellow boatmen valued Lint's juvenile muscle; 1920s river running, with its ankle-breaking portages and lining, was grueling work.

Boatman, Elwyn Blake

The second "husky youngster"[64] Birdseye chose as a boatman was Henry Elwyn Blake. Born in 1896, Elwyn was the oldest of ten children. The family moved in 1909 to Elgin, Utah, a small community opposite the river from the larger town of Green River. Elwyn sometimes joined his father on his attempts at commercial river freighting, but more often he assisted with farming chores. The Blake family moved to Monticello in 1917, where Elwyn, after serving in World War I, found work in farming and at the local newspaper where his father was editor. He was lean and sinewy, with a weathered face that looked far older than its years, and he possessed a reliability that kept him employed. Bert Loper, whom Blake knew from stints as a placer miner, recommended Blake for the 1921 San Juan survey, and he subsequently earned a spot on the 1922 Green River expedition.[65] Blake, in spite of a serious bearing and style that extended to the expression in his photographs, was not without a sense of humor. When Edith Kolb asked the expedition participants to sign their names and occupations in her diary, Blake

---

61. "Weiser Giant Starts Thrilling Trip by Rowboat Down Snake River to Lewiston; Has Made Many Journeys," *Evening Capital News*, December 13, 1922.
62. Ibid.
63. Personal History Form, June 16, 1924, Lint Personnel File.
64. Birdseye, "Surveying the Colorado Grand Canyon," 22.
65. See Westwood, *Rough-Water Man*, for more details of Blake's life.

H. Elwyn Blake in 1923 at the South Rim, Grand Canyon.

Photographer unknown, courtesy of George Lint.

wrote: "Am crosseyed, red whiskered, and ill tempered from arguing religion with Mormons."[66]

Chief Topographic Assistant, Herman Stabler

Birdseye probably spent a good deal of time discussing the expedition members with his colleague Herman Stabler.[67] Stabler was born in 1879 in Sandy Spring, Maryland, the same year that USGS began. At Earlham College, he excelled in athletics as a state champion pole vaulter and became interested in civil engineering. After graduating in 1899, he studied engineering at George Washington University and taught correspondence courses in math and engineering. In 1905, two years after being hired by USGS as a hydrographic aid, Stabler married Bertha Buhler. His early work concerned water pollution, and he also spent a year and a half as an irrigation engineer with the Reclamation Service. Stabler, like Birdseye, was rapidly promoted, and he was known for his intuition, quick mind, belief in justice, and "a rare capacity for lucid expression."[68]

---

66. Edith Kolb diary, entry for July 19, 1923, box 14, folder 1756, Kolb Collection.
67. See William G. Hoyt, "Memoir of Herman Stabler," *Transactions of the American Society of Civil Engineers* 108 (1943): 1641–45, for a more complete biography of Stabler.
68. Hoyt, "Memoir of Herman Stabler," 1644.

Engineer Herman Stabler, chief of the Land Classification Board of the USGS, joined the expedition at Bright Angel Creek.

Emery Kolb Collection, NAU.PH.568-5165, courtesy of the Cline Library, Northern Arizona University.

From 1912 to 1922, Stabler's work centered upon land classification studies related to public domain water management in the West. In 1922, Stabler was made chief of the Land Classification Branch of USGS. He was intensely familiar with the challenges of Colorado River control; before the Federal Water Power Act, the Land Classification Branch oversaw some of the duties transferred to the Federal Power Commission, and Stabler had helped draft the Act's regulations.[69] He also oversaw many of the Colorado River Basin investigations, "outlining plans for effective conservation of the water resources of the entire basin."[70] Stabler hiked in at Bright Angel Creek rather than launching with the party at Lee's Ferry, and in so doing was only along for about two-thirds of the expedition.

Hydraulic Engineer, Eugene C. La Rue

Birdseye and Stabler worked closely with USGS Chief Hydraulic Engineer Nathan Grover to plan the Grand Canyon work. In a memorandum dated February 15, 1923, Grover and Stabler carefully outlined to Birdseye the survey work required in the various reaches of the Colorado River. They specified that "a hydraulic

---

69. Follansbee, *A History*, vol. 2.
70. Hoyt, "Memoir of Herman Stabler," 1643.

Hydraulic engineer Eugene C. La Rue measuring the discharge at Nankoweap Creek, August 12, 1923.

E. C. La Rue 392, courtesy of the U.S. Geological Survey Photographic Library.

engineer, probably Mr. E. C. La Rue, will be attached to the party. His duties will be to measure the flow of tributary streams; examine and photograph possible dam sites; designate sites for detailed survey, outlining the extent and character of surveys to be made; advise as to the necessity of carrying contours out on tributary streams; and in general to advise the chief of party as to the necessity of any feature of the work from the standpoint of power and reservoir site requirements and to assist in the work of the party in any way that the party chief may deem advisable."[71]

Eugene Clyde La Rue was born in 1879 in Riverside, California. His father had tried his hand at both California mining and Indiana farming before finally settling his family, in 1876, in Riverside. The family, with eight children, was of modest means, although it enjoyed a solid reputation within the community. La Rue's father planted his forty acres with grapevines and orange trees, and Rex, as the family called Eugene, grew up amidst and aware of the challenges of irrigation and agriculture.[72] La Rue joined USGS in 1904,

---

71. Nathan Grover and Herman Stabler to Chief Topographic Engineer (Claude Birdseye), February 18, 1923, box 2, folder 1, La Rue Collection.
72. *An Illustrated History of Southern California* (Chicago: Lewis Publishing Company, 1890), 671–72.

immediately after his graduation from the University of California at Berkeley with a Bachelor of Science degree in Civil Engineering.

La Rue spent his first years with USGS as an engineering aide for the Reclamation Service, working on stream gaging in California and Idaho. Tall and lanky, he performed his duties with sufficient competence to be bumped up the career ladder. He married Mabel Elton,[73] and the couple had three daughters. In 1911, he was promoted to the position of hydraulic engineer in the nascent Division of Water Utilization, making field examinations and preparing reports on irrigation, water power, and railroad projects. The following year, he began his Colorado River basin research, which included extensive interviews with individuals and firms concerned with the irrigation in the Imperial Valley.[74]

USGS applauded La Rue's 1916 publication, *Colorado River and its Utilization*. "Mr. LaRue excels in the preparation of reports on water utilization, his report on the Colorado River ... having excited interest even outside the United States because of its originality of treatment and care and accurcy [*sic*] of presentation."[75] A 1920 promotion letter noted: "the most difficult assignments are reserved for him and his reports serve as the basis for action in many of the important decisions made by the Department."[76] As previously discussed, La Rue acted as chairman of the Arizona Engineering Commission in 1922, work that found him determining the quantity of Arizona land suitable for Colorado River irrigation. It also placed him squarely on Arizona's side in its conflict with the other basin states, including California.

Although USGS clearly valued La Rue's talents, there are hints of cantankerousness early in his career. For example, in 1912, Chief Hydrographer M. O. Leighton sent La Rue a written reprimand for whining about sharing a stenographer.[77] In the early 1920s, La Rue seemed to have become even more gruff. Perhaps his irascibility increased as the high dam controversy heated up; perhaps it had to do with his reputed case of ulcers; or perhaps it was because he was in "a heck of a fix financially"[78] and forced to borrow money in order to pay the bills. He did not make it widely known, but

---

73. It is not clear if "Elton" is Mabel Elton La Rue's maiden name or middle name.
74. Eugene La Rue to Nathan Grover, January 29, 1914, La Rue Personnel File.
75. Promotion letter signed by George Smith, June 1919, La Rue Personnel File.
76. Promotion letter signed by George Smith, June 1920, La Rue Personnel File.
77. M. O. Leighton to Eugene La Rue, August 16, 1912, La Rue Personnel File.
78. Eugene La Rue to Scott La Rue, January 20, 1922, box 1, folder 13, La Rue Collection.

Geologist
Raymond C. Moore,
Grand Canyon, 1923.

*Emery Kolb Collection, NAU.
PH.568-3223, courtesy of
the Cline Library, Northern
Arizona University.*

he was simultaneously afraid of water and a fair swimmer.[79] As the launch date approached in 1923, La Rue's mood undoubtedly was not improved by his fears, which he ominously confided in a letter to his brother: "This is a dangerous trip and it is possible that some of us may fail to answer the roll-call."[80]

Geologist, Raymond Moore

The position of expedition geologist proved to be the last one filled. The geologist would "accompany the party to advise the hydraulic engineer and report as to the structural suitability of dam sites, to collect general geologic data, and to assist in the work of the party as desired by the party chief."[81] Geologist Sidney Paige, who had accompanied La Rue on the Chenoweth expedition, commented, "I was offered a place in the Grand Canyon survey of 1923 but turned it down. I told the officials that if La Rue also went only

---

79. Otis Marston, interview with Mrs. E. C. LaRue at Pasadena, November 20, 1948, box 114, folder 4, Marston Collection.
80. Eugene La Rue to Scott La Rue, May 15, 1923, box 2, folder 1, La Rue Collection.
81. Nathan Grover and Herman Stabler to Claude Birdseye, February 15, 1923, box 2, folder 1, La Rue Collection.

one of us would come out. He is sort of an ass."[82] Other prominent geologists with experience in the region turned the job down. Herbert E. Gregory was unavailable; Levi Noble was either unavailable or rejected the offer; and Hugh Miser reportedly "had had enough of rivers" after the San Juan trip in 1921.[83]

Raymond Cecil Moore, a stratigrapher and invertebrate paleontologist, ultimately accepted the position. The son of a Baptist minister, Moore was born in 1892 in Roslyn, Washington, but he grew up in the Midwest In 1913, he graduated Phi Beta Kappa from Denison University, where he also ran track. He promptly found work with USGS as a geologic aide. In 1916, he was awarded his Ph.D. summa cum laude from the University of Chicago, joined the faculty of the University of Kansas as an assistant professor, and became the state geologist for the Kansas State Geological Survey. The following year he married Georgine Watters. In 1919, he was promoted to full professor at the university and was appointed chairman of the Department of Geology the following year. Moore returned temporarily to USGS in 1921, when the Kansas survey's appropriation was not included in the state's budget.[84] A well-published scientist, Moore also spoke six languages; he was insightful, confident, artistic, driven, demanding both of himself and those around him; and was intolerant of incompetence.[85]

Topographer, Roland Burchard

Birdseye was an accomplished topographer, but he needed a second topographer to round out the scientific crew. Roland Whitman Burchard, who was reserved, fit, dependable, and meticulous, took the job.[86] The thirty-seven-year-old Burchard was born near Abilene, Texas. His parents ran what began as a sheep ranch and later became

---

82. Sidney Paige, interview with Otis Marston, August 7, 1964, box 114, folder 4, Marston Collection.
83. Raymond C. Moore, interview with Otis Marston, May 1948, box 152, folder 29, Marston Collection.
84. Philip S. Smith, Memorandum for the Secretary, May 19, 1921, Moore Personnel File.
85. Daniel F. Merriam, "Rock Stars—Raymond Cecil Moore," *GSA Today* 13 (2003): 16–18; Carl O. Dunbar, "Raymond Cecil Moore," in *Essays in Paleontology & Stratigraphy: R. C. Moore Commemorative Volume*, ed. Curt Teichert and Ellis L. Yochelson (Lawrence: University of Kansas Press, 1967), 5–7; Rex C. Buchanan, "'To bring together, correlate, and preserve'—a history of the Kansas Geological Survey, 1864–1989," *Kansas Geological Society Bulletin* 227 (February 2003), chaps. 7–8,, http://www.kgs.ku.edu/Publications/Bulletins/227/index.html (accessed September 29, 2004).
86. Our information on Burchard was drawn from Roland Whitman Burchard, "A Life Sketch by Roland Whitman Burchard," 1967, unpublished manuscript in the possession of Preston Burchard; and the Burchard Personnel File, National Personnel Records Center.

Topographic engineer Roland W. Burchard at Diamond Creek in early October.

Emery Kolb Collection, NAU.PH.568-5154, courtesy of the Cline Library, Northern Arizona University.

a dairy and cotton farm, all marginal endeavors. His mother died at an early age, and Burchard was raised by his sister Kate. He spent three years assisting a cousin with a fledgling telephone company, then enrolled in Austin College, with his eye on eventually becoming the company's general manager. By the time he graduated, the telephone company had folded and he needed another career.

Burchard's cousins, who lived in the District of Columbia, urged him to apply for work with the Census Bureau. He got the job and worked as a clerk from 1910 to 1912. In 1914, he joined USGS as a topographic aide. His first assignment was in Mohave, California. Burchard described this job positively: "I ... started my education as a topographer and liked it. I liked riding and packing horses and living and working in the wide-open spaces ... in fact, I liked the U.S.G.S."[87] After a brief stint with the Army Engineers as a second lieutenant mapping stateside training camps during World War I, Burchard returned to his USGS position as a topographer.

A 1920 promotion letter noted "the results of his [Burchard's] work ... are of the highest grade and complete in every respect.

---

87. Burchard, "Life Sketch," 3.

He has had independent charge of difficult assignments and his ... ability to expedite and finish his work economically should be recognized ... His chief [feels] assured that he will get results if any one will."[88] This glowing praise referred to Burchard's work on the lower Colorado River, where he mapped the Boulder Canyon dam site while drilling crews tested the surrounding rock. He also went upstream by boat "carrying a river traverse as far as conditions of stream and terrain would allow"[89] (Last Chance Rapids), where he installed a benchmark and cairn. Returning to camp downstream, he received a letter stating that the Boulder Canyon site had been eliminated from consideration and that he was to proceed about five miles downstream from that site to Black Canyon. There, Burchard identified what he thought the best dam site and mapped it, and his choice became the future site of Hoover Dam. Burchard completed his survey work at Bulls Head Rock, near present-day Davis Dam.

While conducting the Boulder Dam surveys, Burchard was courting a Texas belle named Eleanor Coleman, who came from a somewhat higher social class. As he described it, "I admit that being borrowed by the U.S.R.S. and mapping such an important project as the 'Boulder' dam had given my self-confidence quite a lift and selling the farm ... was giving my finances quite a lift also and I had decided I could no longer get along without that Gal and I told her so."[90] They married in 1921. Later that year, Burchard was promoted to topographic engineer.

Rodman, Frank Dodge

For a rodman, Birdseye chose Francis (Frank) Beverly Dodge,[91] whom he first met when engaged in topographic work in Hawaii. Birdseye undoubtedly was familiar with one of Dodge's most useful river skills: he was a phenomenal swimmer. He honed his swimming technique in the Pacific waters surrounding Hawaii, where he was born in 1891. His father, a civil engineer and a graduate of MIT,[92] served as a chief assistant on the Hawaiian Government Survey and was superintendent of the Bishop estate, which included the Bishop

---

88. Promotion letter signed by George Smith, June 1920, Burchard Personnel File.
89. Roland Burchard to L. Don Harris, December 18, 1952, box 25, folder 17, Marston Collection.
90. Burchard, "Life Sketch," 5.
91. Dodge, *Saga*; Dodge Personnel File; Reilly, *Lee's Ferry*.
92. "Alumni Notes," The Tech 18 (November 6, 1898): 83, http://www-tech.mit.edu/archives/VOL_018/TECH_V018_S0187_P009.pdf (accessed March 13, 2006).

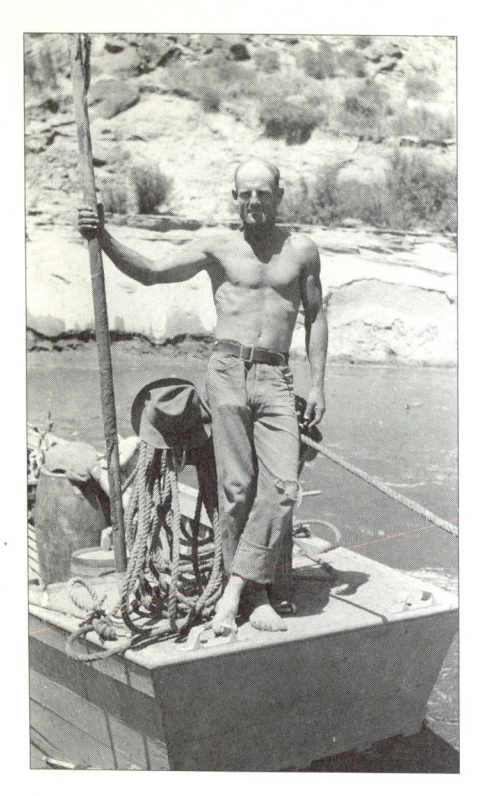

Frank B. Dodge at Glen Canyon, October 8, 1921.

E.C. LaRue 247, courtesy of the U.S. Geological Survey Photographhic Library

Museum and Kamehameha School. Expectations were high for young Frank, but he left Punahou Prep prior to graduating, and went to sea as a "boy before the mast" on a barque taking sugar to New York. In 1909, he returned to Hawaii and worked as a stream gager on a plantation. When USGS took over the gaging station the following year, Dodge became a federal-government employee.

In spite of Dodge's inclination to change jobs frequently, Birdseye and others saw his potential and urged him to continue his education at his father's alma mater. Dodge boarded a boat bound for California, carrying funds intended for his schooling. He got as far as San Francisco, where he squandered the money at night spots, developing a taste for liquor that would haunt him the rest of his life. Stuck on the continent, he found temporary work with USGS as a recorder. Dodge drifted between the mainland and Hawaii for the next few years, before enlisting in the Navy during World War I. Afterwards, he moved to Arizona, where he worked as a carpenter, cowpuncher, and Forest Service aide. In 1921, USGS again hired him to work with the Southern California Edison parties at Lee's Ferry, working with, among others, La Rue. Dodge

Frank E. Word was the cook on the trip as far as Havasu Creek, where he hiked out. The initials "F.W." are printed on his lifejacket.

Emery Kolb Collection, NAU. PH.568-807, courtesy of the Cline Library, Northern Arizona University. Detail view.

Cook Felix Kominsky at an unknown location in western Grand Canyon.

Emery Kolb Collection, NAU.PH.568-3333, courtesy of the Cline Library, Northern Arizona University.

was strong and lean on his five-foot-ten-inch frame; he was versatile and not overly assertive. In Edith Kolb's diary, Dodge noted in addition to his name and address that he was "Young, handsome and bald headed. Weight 160 lbs. including my whiskers."[93]

Cooks, Frank Word and Felix Kominsky

At La Rue's recommendation, Birdseye hired Frank M. Word of Los Angeles as the expedition cook. Perhaps in his forties, his photographs show a look of perpetual anxiety. Word is something of a mystery; his name appears infrequently in written records of the expedition, perhaps because he left the expedition before it was completed. La Rue regarded him highly enough that when he learned what the salaries were for the boatmen, he asked that Birdseye increase Word's pay from $125 to $150 per month because "he is a good man."[94] Word was an experienced cook, as indicated by his stated occupation in city directories of the time.[95]

Little more is known of Felix Kominsky, a replacement cook who joined the trip at Havasu Creek. Kominsky is perhaps best known for his rather jovial appearance in photographs, which have been widely published.[96] According to Birdseye, "he is a Pole and his real name is Kominsky, but a recruiting officer in the Army told him an Army cook did not need such a long name and changed it to Koms. Felix has used this handle ever since. Felix was born in a coal-mining town in Pennsylvania and has not yet learned to speak English clearly; also can hardly write. He is, however, a good cook - fat, jolly and clean and not afraid of water or work."[97] Kominsky was Fred Harvey House trained,[98] which is noteworthy information given that the 1920s were the heyday of the Harvey House restaurant dynasty.

---

93. Edith Kolb, diary, box 14, folder 1756, Kolb Collection.
94. Eugene La Rue to Claude Birdseye, March 23, 1923, box 2, folder 1, La Rue Collection.
95. Los Angeles City Directory (Los Angeles: The Los Angeles Directory Company, 1924): 2394.
96. The best known of the photographs of Kominsky, made classic by Buzz Belknap's 1969 river guide to Grand Canyon (among other publications), shows him sharpening a very large knife while flanked by Blake and Lint. Buzz Belknap, *Grand Canyon River Guide* (Salt Lake City, UT: Canyonlands Press, 1969). See photo on page 255 below.
97. Birdseye, diary entry for September 14, 1923.
98. Claude H. Birdseye, "Boating in the Rapids of the Grand Canyon," *The Army and Navy Courier* (February–March 1926): 40.

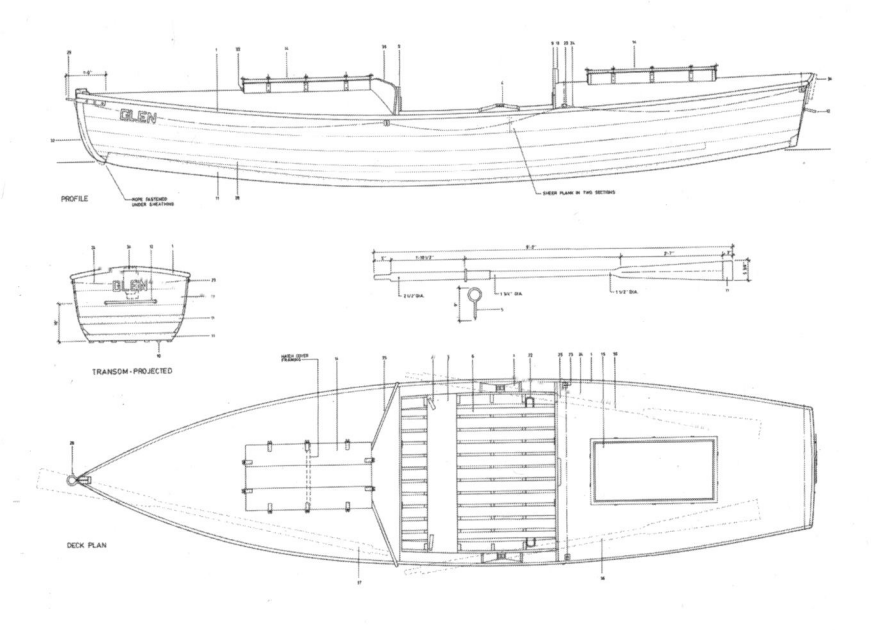

Engineering plans of the *Glen*.

From drawings by Todd Bloch, courtesy of the Historic American Engineering Record Collection, National Park Service.

The Equipment

Boats

From Washington, Claude Birdseye organized the complex logistics of assembling and conducting a Grand Canyon expedition. He needed four main boats for the expedition, and three were already in storage at Lee's Ferry. Southern California Edison had commissioned these boats, which were used during the 1921 Cataract Canyon expedition. At least two of them had been built by Fellows and Stewart Shipbuilding Works on Mormon Island, in San Pedro Harbor outside of Los Angeles, based on a design honed by Nathaniel Galloway. Two of the boats stored at Lee's Ferry were beefier versions of the original Galloway design, measuring eighteen feet long and weighing nearly eight hundred pounds empty. The third, which the 1923 crew renamed the *Boulder*, was sixteen feet long and of a slightly different design. Birdseye contracted with Fellows and Stewart to build a fourth boat, another eighteen-footer. Freeman and Kolb, who was staying in Los Angeles at the time, served as consultants on its construction. Fitted with watertight compartments made of tin, the boats could haul large amounts of gear.[99] Birdseye also purchased a fifth craft: a King 14-Foot Special folding-canvas canoe that Dodge would use both in his rodman duties and as a safety boat in the event of an accident.[100]

To accommodate the limited storage space in the boats, and because he realized many rapids would have to be portaged, Birdseye determined that the gear to be used on the expedition had to be compact, portable, and as waterproof as possible. Dunnage would be placed in rubber bags, and cameras, films, maps, and survey gear would be in watertight boxes. Birdseye drew heavily upon Kolb's previous experiences for details: the proper oar length, the contents of the repair kit, the length and type of rope, and whether or not each man should have a life preserver.[101] In spite of his efforts to keep the load to a minimum, Birdseye's expedition would literally carry tons of gear through Grand Canyon, a serious liability when the plan was to line or portage most of the serious rapids.

Surveying, Cameras, and Other Scientific Equipment

Birdseye, who would later write a manual on how to conduct topographic surveys,[102] carefully designed the survey equipment. He wrote that "specially constructed plane-table and telescopic alidades were used, distances, of course, being measured by stadia. The alidade was of the micrometer type [possessing fine-adjustment capability], permitting measurements of long distances in cases where instrument and rod stations were too far apart for accurate stadia determinations."[103] Burchard ordered a fourteen-foot long, folding stadia rod made to his specifications; this stadia rod could be read from 2800 feet away, allowing for long survey shots from one instrument station.[104] Other survey equipment included three additional alidades of varying styles (sight, Bumstead, and Gale), extra stadia

---

99. Handwritten notes on what appear to be rough designs for the compartments are in NARA, Record Group 57, Records of the Topographic Division.
100. "King Folding Canvas Boat Co., Kalamazoo, Michigan, U.S.A." catalog, NARA, Record Group 57, Records of the Topographic Division. The price of the canoe is given as eighty dollars, with an additional skin costing thirty dollars. Its weight is listed as 100 pounds, with a carrying capacity of 1,200 pounds.
101. Claude Birdseye to Emery Kolb, March 7, 1923, box 6, folder 666, Kolb Collection; Claude Birdseye to Emery Kolb, April 17, 1923, box 6, folder 665, Kolb Collection.
102. Claude H. Birdseye, *Topographic Instructions of the United States Geological Survey* (Washington, DC: U.S. Government Printing Office, 1928).
103. Birdseye, "Surveying the Colorado Grand Canyon," 23.
104. Watkins, *The 1923 Surveying Expedition*.

rods, an aneroid barometer, tripod, four Brunton compasses, levels, tapes, field glasses, and field books, all of which needed to be protected against water and sediment.[105]

La Rue would be the official photographer for the expedition, a decision that caused irritation with Kolb before, during, and after the trip. La Rue brought three still cameras[106]—a panoramic camera,[107] a normal format "3A" camera,[108] and a stereo camera—and a hand-cranked Debrie Sept movie camera.[109] The party needed a variety of equipment for La Rue and Moore, including a tool kit, map paper, a current meter for stream gaging, and rock hammers for geologic investigation. Finally, Stabler compiled a sort of river guide from the journals and publications of previous trips, with notes on rapids.[110]

Food, the Resupplies, and the Radio

Birdseye knew they could not possibly carry enough food for the trip, which was projected to last about eighty days given its ambitious agenda. He delegated to Frank Word the preparation of the menu and acquisition of cooking gear.[111] The official ration list for ten men for thirty days included 240 pounds of flour, 300 pounds of rice, 100 pounds of sugar, 100 pounds of potatoes, 100 pounds of bacon, 100 pounds of cornmeal, 100 pounds of ship biscuit, 72 cans of Carnation milk, 100 pounds of butter, and 100 pounds of lard, as well as canned vegetables and fish, dried fruit, beans, baking supplies, spices, coffee, tea, cocoa, and dried

---

105. See "Supplies: C. H. Birdseye, Flagstaff, Arizona" for a listing of trip gear. NARA, Record Group 57, Records of the Topographic Division.
106. In his photographic notebook, La Rue recorded numbers, dates, and subjects for all his images, sorted by camera type. He also included detailed sketches of some camera locations. A photograph of La Rue's wife, Mabel, is pasted inside the front cover. La Rue photographic notebook, box 10, folder 5, La Rue Collection.
107. La Rue's panorama camera was probably made by Al-Vista, a Burlington, Wisconsin firm that made at least eighteen different models. See Bill McBride, "Al-Vista Panoramic Cameras," Santa Barbara, CA, March 1989.
108. La Rue's 3A was likely one of the many Kodak cameras that contained that model number. Kolb also brought a 3A camera, but we do not know if it was the same model as La Rue's.
109. The Debrie Sept was a four-pound, spring-motor drive camera capable of taking both stills and short movie clips on 35 mm roll film. A copy of the original brochure is held at NARA, Record Group 57, Records of the Topographic Division.
110. A copy of what is presumably the river guide prepared by Herman Stabler is in NARA, Record Group 57, Records of the Topographic Division. It includes summaries and notes by river mile taken from Frederick Dellenbaugh's *A Canyon Voyage*, Stanton's expedition, and the Kolb brothers' trip.
111. Claude Birdseye to Eugene La Rue, March 16, 1923, box 2, folder 1, La Rue Collection.

eggs.[112] In order to avoid the culinary fate of the first Powell expedition, which had to subsist on coffee, moldy flour, and rancid bacon towards the bitter end, the 1923 USGS expedition would be resupplied at points where land parties carrying substantial goods could reach the river. To handle the logistics of transportation and resupply, Birdseye's cousin Roger hired Charles Fisk, who worked for Grand Canyon National Park, to manage the animals and determine the resupply points.[113]

To the growing pile of dunnage Birdseye added one important innovation for a river trip: he purchased a radio. Radio broadcasting began to claim a foothold in America in 1920, and by 1923, the number of stations had greatly increased. Radios were modern, and modernity was a buzzword in the 1920s.[114] To bring a radio along on a Grand Canyon river expedition was not only to have news of the outside world, it was news in and of itself. This move, to take the first radio on such an expedition, was carefully designed for publicity purposes as well as a limited means of one-way communication from the outside world to the expedition.

A powerful new radio station, KHJ, owned and operated by the *Los Angeles Times*, was more than willing to make special broadcasts to and about the expedition.[115] It was a public relations coup for both sides. Because both La Rue and the publishers opposed the Boulder Dam proposal and Arthur Davis, the *Times* backed USGS work, which was paid for by Southern California Edison. The *Times* frequently and favorably reported La Rue's work. Freeman, who came from Southern California, also may have had friends at the paper. La Rue communicated directly with KHJ manager, John S. Daggett,[116] known as "Uncle John" to thousands of devoted

---

112. "Ration List – Colorado River Trip. 10 men for 30 days (300 rations)," NARA, Record Group 57, Records of the Topographic Division.
113. Charles Fisk was apparently not happy that in the torrent of publicity about the expedition, Roger Birdseye received the credit, while he (Fisk) did the actual work of packing. Michael Harrison to Otis Marston, January 17, 1970, box 69, folder 3, Marston Collection.
114. Timothy D. Taylor, "Music and the rise of radio in 1920s America: Technological imperialism, socialization, and the transformation of intimacy," *Historical Journal of Film, Radio and Television* 22 (2002): 425–43.
115. "KHJ Liaison in Colorado River Work: Keeps Men on Dangerous Survey Expedition in Touch with Government," *Los Angeles Times*, August 5, 1923 (accessed March 1, 2006).
116. John Daggett had been working as a reporter for the *Los Angeles Times* when KHJ summoned him to manage the station, which went on the air late in 1922. He held the position until November 1927, when the station was sold. "Radio Bids 'Uncle John' Adieu," *Los Angeles Times*, November 30, 1927 (accessed February 28, 2006). An earlier issue of the paper reported, "Two things have established Uncle John in the hearts of his

listeners. La Rue's response to Birdseye's instruction read "I will prepare a statement for Mr. Daggett, and will give him considerable information regarding the character of the Canyon of the Colorado, the purpose of our trip, the character of the surveys and the importance of this work in connection with the Colorado River project."[117] The *Deseret News* of Salt Lake City was also apparently enlisted as a broadcaster, although the men never noted special bulletins coming from them.[118]

Two weeks before the expedition launched, an article appeared in *The Literary Digest* reporting that radios would not be able to receive signals in a "walled-in space." This claim only heightened interest in the expedition's use of a radio in a deep canyon, the ultimate walled-in space.[119] Birdseye prepared accordingly: "A Westinghouse Aeriola Sr. receiving set was remodeled and amplified, using U.V. 201-A tubes throughout. Wet batteries could not be recharged, so dry cells only were carried—four 1½-volt dry cells being used to light the filaments and a small 22½-volt B battery to charge the plate. The complete set, including battery and phones, was packed in a water-tight box weighing about 20 lbs."[120]

All this equipment made the 1923 USGS expedition heavily laden, which would affect their ability to avoid running rapids. With their four 800-pound boats and estimated 2 tons of equipment,[121] they faced the choice of running difficult or treacherous rapids or carrying their gear around them. Late in the trip, this decision—to run or portage—would cause acrimony among the boatmen, pitting the "husky youngsters"—who at that point, had learned to run rapids—versus the more cautious Kolb and Birdseye.

The Infighting Begins

Birdseye was a scientist, first and foremost, but his long field experience undoubtedly made him a leader of men. His prime objective for the 1923 expedition was to conduct the necessary surveys and

---

listeners—his strong, sympathetic voice, and his spontaneous, hearty laugh." Ben A. Markson, "Praise Where it is Due," *Los Angeles Times*, April 8, 1923 (accessed February 28, 2006).

117. Letter dated March 23, 1923, La Rue Collection.
118. "Department of the Interior, Memorandum for the Press: Surveys of Grand Canyon of the Colorado," n.d. NARA, Record Group 57, Records of the Topographic Division.
119. "How Far Can My Radio Set Receive?" *Literary Digest* 78 (July 14, 1923): 25.
120. Birdseye, "Radio in the Grand Canyon."
121. The estimate of 2 tons of equipment comes from a caption on E. C. La Rue photograph 445 (see photo on page 227 below), U.S. Geological Survey Photograph Library.

studies, and to do them well. He was a capable administrator, as evidenced by his rapid promotions during his military and USGS careers. While he had never conducted a river trip, he took the dangers of river running seriously, as one would expect from a man experienced both in war and the outdoors. The crew he had assembled, while capable, was laden with strong and divergent personalities. Birdseye knew that his most important communications had to be with his head boatman, Kolb, who had one of the strongest personalities of the men he had hired.

In March 1923, Birdseye wrote to Kolb that "one of the prime requirements on a trip such as the one we are planning is that everyone will be in perfect harmony and all working toward one common end. I think we will have a congenial party and look forward to the trip with considerable pleasure, although I know much of the work will be hard and many of the experiences unpleasant."[122] He must have suspected that some of those unpleasant experiences would involve interpersonal conflict, particularly involving Kolb and La Rue.

In the same letter, Birdseye continued, "I hope you did not resent any of the statements in my telegrams regarding my objections to your taking a moving picture outfit, but I am very keen on making this trip a success from an engineering point of view and feel that I must disregard all other considerations."[123] But the issue still bothered Kolb, and continued to do so. Other anxieties plagued Kolb as well, including his daughter Edith's taking ill with a malady sufficiently serious to warrant a trip to Los Angeles for treatment a few months before the proposed launch date. While there in May, Birdseye and Kolb met in person and further discussed the situation. On June 3, Kolb wrote Birdseye, "if you knew the true facts of the tremendous struggle throughout the past 20 years which my brother and I have had to contend with for our existence and the right of our business at the Canyon, I believe you would not criticize me in my first request, nor when not granted in the last message, in which I asked you to stipulate a salary with the exclusion of a motion picture camera on the trip. I feel it would be self extermination for me to take part in a trip where others obtain motion pictures as it is in such pictures, more than anything else, we make our living; and I am confident that you or your Department of the Government would not knowingly want to injure us."[124]

---

122. Claude Birdseye to Emery Kolb, March 7, 1923, box 6, folder 666, Kolb Collection.
123. Ibid.
124. Emery Kolb to Claude Birdseye, June 3, 1923, box 6, folder 683, Kolb Collection.

Kolb assured Birdseye that if he were permitted to take a movie camera, it would not interfere with his duties as head boatman, and that he would like the opportunity to provide newsreel footage for commercial release. Birdseye's five-page reply was cordial and to the point. "I think it is advisable that you and I have a definite understanding before starting our trip and for that reason will place in this letter all our plans or proposals regarding pictures taken on our canyon trip." He expressed his sympathy for Kolb's business dilemma, while noting that USGS "can not in any way restrict its activities or records in favor of any individual or organization."[125]

Birdseye outlined his plan for photography. Any of the crew members could bring still cameras; La Rue, Freeman, Blake, and Moore all had them. Kolb was free to bring a small movie camera, and USGS would provide film if he gave USGS a copy of the footage. Ellsworth Kolb was welcome to join Fisk on resupply trips in order to photograph the expedition. La Rue would operate the official motion-picture camera, a Reclamation Service photographer might shoot some footage here and there, and Fox Movietone News had made arrangements to take footage at trail-accessible points along the river. Kolb was livid—he wanted a monopoly on film rights to Grand Canyon.

Birdseye, clearly annoyed, informed Kolb that the salary he offered Kolb—$500 per month—was greater salary than his own $415 per month, and was equaled only by that paid the director of USGS. He reminded Kolb of his earlier stipulation that USGS priorities trumped private interests and expressed his belief that this, his final offer, was fair. He wrote "you must decide for yourself whether or not you care to be a member of the party. If you can not subscribe heartily to my proposals and agree to help us in every possible way I shall be compelled to withdraw my offer and get along without your services, much as we desire them and need them."[126] Kolb decided to join the expedition, but he would not still his objections: he felt that he had been lured by false promises and was cheated out of his self-claimed rights to movies about Grand Canyon.

The Final Preparations and the Journey to Flagstaff

In the months prior to the expedition's departure, Birdseye honed his plans, particularly those concerning assembly of crew

---

125. Claude Birdseye to Emery Kolb, June 11, 1923, box 6, folder 670, Kolb Collection.
126. Ibid.

and gear in Flagstaff, travel to Lee's Ferry, and the work of getting a river expedition on the water and moving downstream. Birdseye had set July 16 as the day the expedition was to assemble at Lee's Ferry. In 1923, Flagstaff was a small railroad and lumbering town on the high plateau of northern Arizona and the closest community of any consequence to the launch point at Lee's Ferry. Roger Birdseye lived in Flagstaff, which would facilitate trip logistics.

Claude Birdseye, accompanied by his son Charles, arrived early in July to work with Roger. He traveled to Grand Canyon to meet Kolb, instructing him "to take full charge of the boats" and to distribute them among the boatmen. When Lint arrived on the Santa Fe railroad on July 14, he found Birdseye, Blake, and Dodge already in town; Word and Freeman arrived the following day and La Rue, Moore, and Kolb arrived the day after. Kolb brought Blanche, Edith, and Edith's friend Catharine Pahl of Los Angeles.[127] Freeman came to Flagstaff via Grand Canyon, where he had spent ten days finishing the final chapter of his forthcoming book. Burchard arrived just in time on the day the party left for Lee's Ferry, having been detained by the birth of Preston, his first child.

In his autobiography, written decades later, Dodge made several observations of the meeting in Flagstaff. He wrote that Freeman was an "explorer, big game hunter, northern riverman, at 257 lbs." and "was recommended through La Rue." According to Dodge, the meeting was jaw-dropping: "one day while I was sitting in the lobby [at the Weatherford Hotel], who should walk in looking like an overweight sportsman dude, but a nattily dressed man in rompers, wool knee stockings, two-tone oxfords, and a cane. Wealth and good living stuck out all over him." He followed this with the observation, "I could remember reading Cosmopolitan short stories that were illustrated with drawings of huge log-riding supermen, tearing through rapids, and having shoulders four feet across and being some eight feet tall. Real supermen of the north, and I thought it would be hard keeping up with such." Despite the first impression, Dodge grew to like and respect Freeman.

Dodge thought that although Freeman "was sailing under a lot of wind, I learned he was a writer, a newspaper correspondent, and if poetic license regarding one's past isn't allowed in that field, what is? He had to blow his own horn to sell his stuff. The more blowing

---

127. Catharine Pahl's father was a physician who was a friend of the Kolbs; he may have been the one treating Edith for her unidentified illness.

the more sales. If to get his 'Big game hunter' title, he rode the 27th elephant while the Prince of Wales rode the first on a tiger hunt in India, still none of us had ever ridden an elephant. And if he exaggerated his Columbia River magazine stories, it was to sell them. In camp he was as honest and unassuming as Birdseye and was always trying to overcome his handicap in weight. Where I'd walk off with a hundred pounds at a portage it was all he could do to carry himself around. And this was the man I'd been afraid of! Yes, he was a misfit in that job but a very pleasant misfit and, since being an intellectual whose mind took to heights above most of ours, he was easy to put up with."[128]

With these lines, Dodge touched upon one of many personality conflicts that splintered the expedition almost immediately. For most of the trip, the boatmen were divided into fractious groups. Freeman was at odds with Kolb, Blake, and Lint. Lint and Blake, at least initially, were loyal to Kolb. Kolb never stopped grousing about the photography, fanning his conflict with Birdseye, extending it to Freeman and La Rue (both of whom had cameras), and ultimately casting a pall over the entire expedition. Burchard, quietly, was more sympathetic to the "working man" and greatly admired Kolb. La Rue was at odds with nearly everyone. Dodge appeared to like Freeman, and later correspondence indicated he was friendly with La Rue.

Birdseye kept his feelings mostly to himself, but he did confide in his head boatman, Kolb. Dodge recognized the significance of this, writing in his autobiography "being strictly a business proposition, it was foolish to take chances [on the river]. A lost boat would have meant the end of the survey. It was here Emery Kolb used his head, and Birdseye gave him his head. That is where Birdseye shone. Too many men hire a head boatman; then when on the river, seem to forget what he was hired for. And Emery was old enough to be able to read of others running a certain rapid and not let it influence his actions of the moment."[129] With the singular action of keeping those communications open, Birdseye justified the faith USGS placed in his leadership.

The work slated for Lee's Ferry before the trip could launch had two components. First and foremost, the stored boats had to be renovated and made watertight. Second, the previous Glen Canyon survey had to be carried into Marble Canyon, the first

---

128. Dodge, *Saga*, 35–36.
129. Dodge, *Saga*, 37.

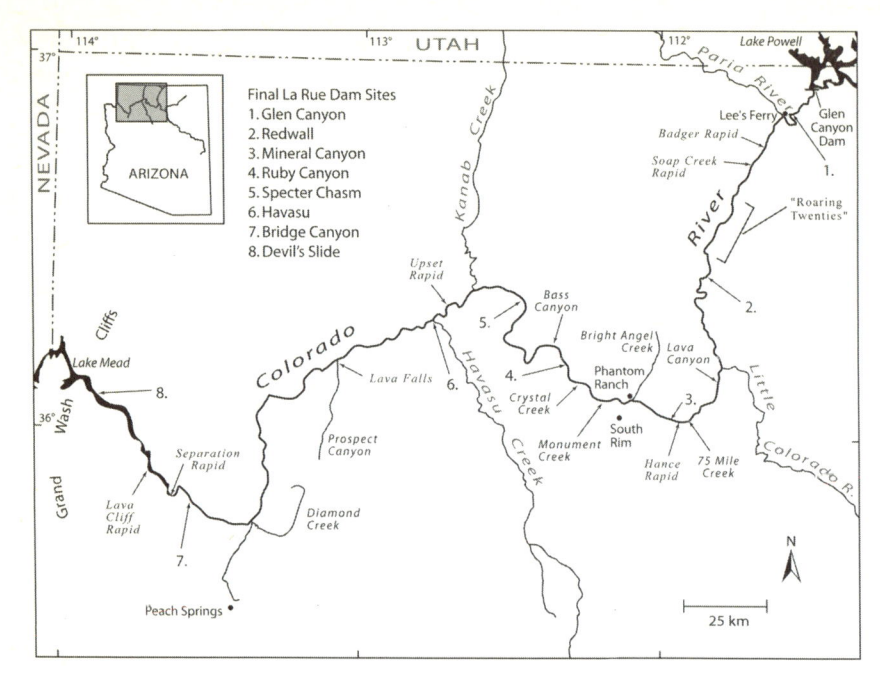

The Colorado River through
Grand Canyon.

reach downstream from Lee's Ferry. In their February 15 memorandum to Birdseye, Grover and Stabler wrote: "From Lee Ferry to the mouth of Badger Creek, a distance of about 7 miles, the survey and map should be essentially a continuation of the mapping of Glen Canyon on a scale of 1:31,680 and with contour intervals of 20 feet on land and 5 feet on water surfaces. Topography in this section should be run up to elevation 3600 and a detailed dam-site survey may be required near the end of this stretch."

Birdseye also developed specific plans for his survey staff during the expedition. Once the party had launched, "From the mouth of Badger Creek to the north boundary of Vishnu Quadrangle, about 53.5 miles from Paria River, the canyon should be mapped up to elevation 3150 on a scale of 1:31,680 and with contour intervals of 50 feet on land and 5 feet on water surfaces. Probably one dam-site survey and possibly more will be required in this section." Furthermore, "detailed dam-site surveys should be made on a scale of 400 feet to the inch and will contour intervals of 10 feet unless a larger scale or smaller interval is specified by the hydraulic engineer.

Such surveys should include an accurate cross section along the axis of the proposed dam and should show sufficient topography to cover construction and operation features. In case existing surveys, when enlarged to the scale for dam sites, will show the topography as satisfactorily as a new survey on the larger scale, the field work may be limited by agreement with the hydraulic engineer."[130]

As his words indicate, Birdseye had ambitious plans for mapping dam sites in Grand Canyon, and those plans would require extensive time and commitment from his surveyors and crew. Before any of those plans could be enacted, Birdseye had a large logistical problem. He needed to get the personnel, the boats, and several tons of equipment and dunnage to Lee's Ferry. It would take a day and a half to travel over a route that now takes three hours by pavement.

---

130. Nathan Grover and Herman Stabler to Claude Birdseye, February 18, 1923, box 2, folder 1, La Rue Collection.

4

A Cumbersome Journey

Flagstaff to Lee's Ferry to the Little Colorado River

On July 18, 1923, Claude Birdseye, the members of his expedition, and an extended traveling party left Flagstaff en route to Lee's Ferry.[1] They had spent several days sorting groceries, packing equipment, and loading the new boat on a truck, well padded to minimize damage during transit over the rough dirt road. They started their trip before noon, taking three trucks and three automobiles to accommodate twenty-three people, equipment, supplies, and the new boat. They camped at Cedar Ridge, roughly halfway to Lee's Ferry. That evening, they enjoyed Frank Word's cooking. All was not jovial on the first night: Kolb described the campfire singing as "stale."

The following day, they had to repair washouts in the road created by summer rains. The truck bearing the boat was back heavy, prompting two men to ride on the radiator to provide balance. The Dugway, the final approach to the ferryboat entry at Lee's Ferry, was steep and winding, and occasionally they had to back up the truck bearing the boat to take the treacherous curves. La Rue and Kolb commenced their cinematic rivalry, both filming the trucks descending the hill. They reached Lee's Ferry at noon and were greeted with blistering heat. As they would do frequently for most

---

1. Our abbreviated discussion of the events preceding the launch of the expedition on August 1 reflects the previous, and considerable, discussion in other publications. In particular, see Westwood, *Rough-Water Man*, 126–32; Lavender, *River Runners*, 59–61; Reilly, *Lee's Ferry*, 301–2.

La Rue took this image with his panorama camera on August 12, noting in his photographic log that the picture was taken "upstream from R.B. [right bank] at mouth of Nankoweap." The view actually is downstream from the mouth by about a quarter of a mile.

E. C. La Rue 393, courtesy of the U.S. Geological Survey Photographic Library.

of the next three months, the crew swam in the Colorado River, which was running at an above-average 34,700 ft^3/s. The crew learned from a USGS colleague stationed at Lee's Ferry that the ferryboat had sunk a few days before, which made transport across the river problematic until the boats were made watertight. They camped on the left (east) bank near the building that housed the three Southern California Edison boats that they were borrowing for the expedition.

The stored boats were dry and cracked, both from their previous trip through Cataract Canyon in 1921, followed by the ensuing long

Repairing the road in Tanner's Wash, en route to Lee's
Ferry, July 18 or 19, 1923.

Lewis Freeman photograph, Emery Kolb Collection, NAU.PH.568-5219, courtesy of the Cline Library, Northern Arizona University.

storage in the desert air. The boatmen commenced the multi-day task of renovating the boats: patching, caulking, painting, and then sinking the boats in the shallows to allow the dry wood to absorb some moisture and close the cracks. This maintenance was necessary to put the boats into serviceable condition for the long expedition. They set up the folding boat as well and tested it, finding it to be of dubious quality and in need of reinforcement. After dinner, some of the group crossed to hydrographer Irwin Cockroft's dwelling and listened to the radio broadcast from Los Angeles. A storm blew in that night, but it only brought much-needed cooling winds. The Kolb family tent blew down, giving Emery a sleepless night.

They arose on July 20 to a rising heat and a dropping river. The topographic crew moved up the Paria River, which enters the Colorado River just downstream from the ferry itself, while the boatmen continued their repairs. They painted one of the boats a deep red. Freeman irritated at least Kolb with his "unwelcome advice." Returning in the afternoon, the topographic team brought back fresh fruit from Lonely Dell Ranch, which was at that time occupied by a group of polygamists, the Johnson family.[2] That afternoon, the two Birdseyes and La Rue drove downstream along the left bank, and La Rue photographed at several points with his panoramic camera. In a warning they would have to heed later, Lint, out for a swim, was flushed downstream by an eddy just offshore. Dodge swam out and pulled him from the river, but Lint was exhausted and unable to eat supper. That night, the thunderstorms and wind returned, again flattening the Kolb family tent and blowing sand into the others' sleeping bags. The experience was repeated the following day, except the evening was clear and the group once again sang, this time with Moore playing a harmonica.

The next day, Sunday, a group drove the Cockroft's car down the right side of the river, inspecting the future location of Navajo Bridge as well as scouting Badger Creek Rapid from the rim. Everyone else took the day off, washing clothes and hair. The two young girls, Edith and Catharine, complained of the heat and blowing sand, but they adjusted. More storms occurred in the evening, but that did not dampen Freeman's enthusiasm for the radio broadcast from Los Angeles: Freeman's appreciative grunts and applause kept others awake.

On Monday, July 23, the river had dropped to 32,400 ft^3/s. Boat maintenance and equipment testing continued as the crew readied for the coming expedition. The boatmen tried out the life preservers—they had two types, cork and kapok (from the kapok tree's fluffy seed covering, commonly used as stuffing for pillows and life jackets)—during the daily swim and found them bulky but buoyant. The radio yielded a time correction for La Rue and the first report of their expedition. The next day, the surveyors and hydrographers fanned out, mapping and inspecting dam sites. The river rose back to a bright red 34,500 ft^3/s in response to thunderstorm runoff and fell back to about 33,600 ft^3/s the following day. One by one, the boats were shoved into the river, with La Rue and Kolb shooting movies, and Edith taking a ride down the bank, waving a bottle of vinegar as a substitute for Prohibition-banned champagne. Kolb expressed paranoia that La Rue's footage, which was to be sent out for processing, would be stolen and shown on movie screens before the expedition was over; no one else believed his concerns were valid. Claude Birdseye and crew completed the preparatory survey work, and Roger Birdseye returned to Flagstaff, only to return again on July 28 with the mail.

---

2. Reilly, *Lee's Ferry*, 288.

The preparations reached their climax from July 26 through 28 as the boats were all floated and named for various canyons of the Colorado River. The new boat became the *Grand*; the three Edison boats were the *Glen* (the red boat), the *Boulder*, the *Marble* (which had been painted green); and the folding boat became the *Mohave*.[3] They made the *Mohave* more rigid and added airbags to improve its chances for survival in the rapids. It was so hot that La Rue's movie film melted while he loaded it midday, and even Edith's friend Catharine Pahl, who had until this time stayed out of the river, went swimming. Thunderstorms came and went, leaving Edith to gush, "the sunset this evening is too wonderful for words. The clouds are beautifully tinged from pale yellow to deep orange, the softened colors of the rocks and cliffs made a wonderful back ground."[4] Around dark on the twenty-ninth, in a harbinger of things to come decades later, four men arrived in two boats from upstream: it was David Rust and one of the first commercial river trips on the Colorado River.[5]

Edith Kolb entertained herself with card games, swimming, and conversation. In particular, she struck up a friendship with Leigh Lint; her diary suggests she may have wanted the relationship to go further. Freeman continued to irritate, making more suggestions and doing less real work, but nonetheless the boat repair was completed, and the boats were loaded and tested. A group photograph was taken next to one of the overturned boats; Dodge was not present because he was drunk, a condition he would be in quite often in the coming years. "If anybody wants to know why I'm not in the group picture as of the *National Geographic* of May 1924, it's because I swam the river on some errand for Birdseye and while talking to Cockroft … about it, he said, 'Frank, want a cup of my fig wine?' I had a cup and then another and then reported back to Birdseye. When Birdseye became three Birdseyes, I hunted a shady spot and

---

3. The *Grand* was the newly constructed boat; the others were renamed from their days in Cataract Canyon in 1921. The *Boulder* was the renamed *Static* (also called the *Tub*); we do not know which of the other boats—the *L.A.* or the *Edison*—became the *Marble* and which was renamed the *Glen*.
4. Edith Kolb diary, entry for July 27, box 14, folder 1756, Kolb Collection.
5. David Rust began to work in Grand Canyon tourism in 1906. One of his first jobs was to supervise the improvement of a trail from the North Rim down to Bright Angel Creek to make it tourist accessible. He built a cableway across the Colorado River at the terminus of his trail, and operated a tented tourist camp, named Rust's Camp, nearby. Michael F. Anderson, *Living at the Edge: Explorers, Exploiters and Settlers of the Grand Canyon Region* (Grand Canyon, AZ: Grand Canyon Association, 1998), 146–47. His first commercial river trip was in 1917; Webb, Belnap, and Weisheit, *Cataract Canyon*, 33.

The crew at Lee's Ferry, posing with the boats prior to refurbishing them. From left to right: Leigh Lint, Elwyn Blake, Frank Word, Claude Birdseye, Raymond Moore, Roland Burchard, Eugene La Rue, Lewis Freeman, and Emery Kolb. Dodge is not in the photograph as he was sleeping off a drunk, leading several articles to report that the expedition included nine men, not ten.

Lewis Freeman photograph, Topography D9, courtesy of the U.S. Geological Survey Photographic Library.

Lee's Ferry, with the Vermillion Cliffs in the distance, July 1923. The structure in the left foreground is the boathouse where three of the boats to be used by the expedition were stored.

Lewis Freeman photograph, Grand Canyon 308, courtesy of the U.S. Geological Survey Photographic Library.

R. C. Moore sketch of the head of Marble Canyon and the Vermilion Cliffs near Lee's Ferry. Moore made his sketch from the top of the Echo Cliffs south of the river.

R. C. Moore diary, p.4, courtesy of the University Archives, Spencer Research Library, University of Kansas Libraries.

Lewis Freeman, Elwyn Blake, Leigh Lint, and Emery Kolb soaking the *Marble* in the river at Lee's Ferry, as part of reconditioning the boats, probably on July 25. After being stored for two years in the arid Arizona air, the wood planks had become extremely dry and needed to be rehydrated.

E. C. La Rue 321, courtesy of the U.S. Geological Survey Photographic Library.

passed out. That was very potent wine or perhaps the heat had something to do with it."[6]

Burchard wrote a letter to his brother before the trip launched that summarized much of the interworkings of the expedition as it was about to get underway.

> We have been camped here since my arrival on the 18th. Lee's Ferry is just like the other places along the Colorado—Bare infiltrated strata Hot and Highly colored Bluff Red black blue dirty gray—all radiating heat—The same dirty water flowing between mud banks—Here and there treacherous dry-looking sand Beaches and over all the same dry sky. We have a tem. of about 111° every day. We are working around the hills preparatory to starting down river much as last time only then it was up river. Mr. Birdseye is in charge and he has a well equipped outfit. However it is a very topheavy organization—Two movie cameras and a geologist and hydrolic engineer beside four expert Boatmen—under the supervision of Mr. Kolb of the Kolb Bros of Grand Canyon. He is a dandy but most of the High brows talk too much and eat too much and don't work enuf We have a radio out fit and listen to Los Angels concerts every night when not too tired The Los A. *Times* broadcasts messages to us at a certain hour every night—we have received several already.
>
> We have spent most of this morning making movies of the outfit. We start tomorrow and have to get to Bright Angel trail about Sept 10—where we get our next mail.—All news from Eleanor and the baby is encouraging but needless to say I hate to be so long out of touch with the world. We have Four boats with water tight compartments and each man has a life preserver. We are apt to need them as there are a lot of bad rapids and smooth walls and poor landing places—I am afraid the work will go very slow and I admit that I do not care how soon the trip ends altho now that I am in it & want to see it thru of course ... We are about to get ready to start so good Bye. You may be very thankful you are not here—it is going to be a hard hot, rough trip.[7]

On August 1, the boats rigged and personnel set, the expedition launched for its journey through the canyons of the Colorado River en route to Needles, California. They first entered Marble

---

6. Dodge, *Saga*, 36.
7. Roland Burchard to Harry Burchard, July 31, 1923, courtesy of Preston Burchard.

Canyon, which extends from Lee's Ferry to the mouth of the Little Colorado River. Kolb would man the *Marble*, while Freeman rowed the *Grand*, Blake and Lint got the *Glen* and *Boulder*, respectively, while Dodge piloted the *Mohave*. Kolb would be solo, except through the Paria Riffle, while Blake took Word and the kitchen, Lint carried Birdseye and Burchard, and Freeman carried Moore and La Rue. The river was running 26,600 ft^3/s, still unusually high for the season.

For the remainder of their journey, we will let the participants speak for themselves, quoting directly from their diaries, letters, and reminiscences. The discharges we report are the daily average flow for the gaging station at Phantom Ranch.

August 1, 1923— 26,600 ft^3/s

BIRDSEYE: Left Lee's Ferry 9 A.M. Ran the Paria Riffle and another large riffle not rough enough to call a rapid and camped that night at the head of Badger Creek Rapids, 7.4 miles below the mouth of the Paria.

BLAKE: Mrs. Cockroft[8] had given Dodge a birthday cake as we were leaving, which was very good and much appreciated by the balance of the party, who shared it with him.

KOLB: I take Edith and Catharine[9] in my boat over the first rapid about 1½ miles all told to the lower dug way. The children are thrilled with the choppy ride. I wait until I see both autos over the dug way then we pull on, running a more violent stretch of water (cow rapid. dead cow) sailing then on a beautiful smooth stretch until about 4 mi below where we lunch. Stopping at Badger Creek about 3 P.M. where we make camp. Freeman & I cross to look it over. The water is tumbling in tremendous waves. One shoot in the center seems possible. La Rue appears to be frightened, has little to say.

FREEMAN: Birdseye began to run the river line just below the Dugway, with Burchard recording and Dodge rodding. The latter is using the canvas boat and appears to find it very useful, at least in the quiet water. At three miles from the Paria we stopped to look

---

8. Margery Cockroft was the wife of hydrographer Irwin; Reilly, *Lee's Ferry*, 283.
9. Edith Kolb and Catharine Pahl were not the first women to ride in a boat through Colorado River whitewater. The first was Ellen "Nellie" Powell Thompson, sister of John Wesley Powell and wife of A. H. Thompson, who rode through the Paria Riffle in the *Cañonita* with members of the second Powell expedition in 1872. Sadie Staker, school teacher at Lee's Ferry, rode through the same riffle in a boat rowed by a smitten Nathaniel Galloway in February 1899; Reilly, *Lee's Ferry*, 29, 173.

over the first riffle of any size in the canyon. Some rocks showed at the surface near the head on the right, but were easily avoidable. Kolb's boat did some lively jumping about but was in no trouble. The others went through more steadily, although I slopped in the top of a wave or two that splashed La Rue and left a little water in the cockpit. Dodge came through in the canvas boat without taking a drop. We lunched about noon at the so-called bridge site, a narrow section with abrupt walls nearly 500 feet high. It occurs near the upper end of a reach close to a mile long. If a bridge is ever built here it will almost surely be a record-breaker for height.[10] After running another small riffle, a half mile pull in quiet water brought in sound of a heavy roar which we knew must be that of Badger Creek Rapid. We came in sight of it twenty minutes later, landing in a sheltered eddy behind three large rocks on the right bank, where a clean sandbar offered an attractive camp. The rapid looked decidedly rough, and from the right bank it was difficult to fix upon a course where a boat could be certain of avoiding rocks all the way through. Finally it was decided that a very narrow V three-quarters of the way to the left bank offered the best course, and this we will probably run in the morning. Kolb says he rather expects that the first big wave below the head will fill his cockpit, and that he will have difficulty in avoiding a very large hole about half way down where the water pours over a large rock. We have also noted another boulder on the left of the V, over which an abrupt waterfall is pouring, and decided that it will have to be passed at very close range. From the left side of the canyon we tossed in driftwood and watched its course into the rapid. There is an eddy making a wide sweep in against the left bank, and outside of this the current sets strongly to the right across the head of the rapid, where it tumbles into a nest of boulders among which it looks as though a boat would have an unpleasantly rough time of it. The course to the V is down the right side of the eddy, pulling steadily against the set of the current to the right. As the V—on account of the abrupt drop of the rapid—is not visible more than fifty to a hundred feet above, the approach will call for a rather nice piece of maneuvering. What will happen below we shall see in the morning. Returning to camp, Kolb told the Colonel that if he cared to risk a boat in the morning he would attempt to run the rapid. If the first boat was not seriously damaged or upset, we would run the others. If it got into trouble, it would probably be

---

10. Navajo Bridge was completed in 1929; Reilly, *Lee's Ferry*, 322–39.

in order to portage. Lining is out of the question, except as a last resort, on account of the tumble of waters over the rocks along the side.[11]

LA RUE: We were viewing the [Badger Creek] rapid from the bank when a thunder storm broke over head. Thunder and lightning was awful. In a few minutes we were soaked. About 15 minutes after the rain started the water began to pour over the canyon walls forming water falls 700 feet high. It was a beautiful sight. After the sky cleared we put up the aerial for the radio. 150 ft of wire leading from the side of the canyon to a post set in a sand bank near the river. It is now 7:15 P.M. We should get the bed time stories from "Uncle John" KHJ at Los Angeles at 7:45.

August 2, 1923 — 24,300 ft³/s

KOLB: Had a fine night's rest. After looking over the rapid from the west side, I take Leigh & Blake across to inspect. Leigh [Kolb meant Freeman] stays on East side with La Rue & Moore. Mr. B. takes movie of each boat[12] as they go through. I go first making my channel perfectly, but the heavy boat diving beneath the waves touches a rock. At the time I scarcely felt it, but later found it had been a sharp pointed rock which punctured my boat 1 x 6 in. leaving water in sufficient to wet the sugar oatmeal, salt and Mr. B's map paper, all of which was supposed to be in water tight sacks and tin case but was not water tight. Leigh ran his boat next. Made a beautiful run, Freeman almost as good. Blake dropped in a hole but did not strike. La Rue helped me empty the boat until Leigh came across. We soon had the hole patched with a piece of my pirate flag, white lead and tin. The bottom is very rotten. Mr. B & Dodge came over and helped dry the maps.

BIRDSEYE: Decided to use celluloid [for the maps] for the rest of the trip.

FREEMAN: Drying out the map sheets and repairing the boat took an hour. The map cases were transferred to the forward hold of the *Grand*, where they will probably remain until I fall into disgrace by puncturing the bottom and letting water in upon them.

11. When Freeman prepared his official, typed account of the expedition, he changed his diary entries as he saw fit. The original entry for August 1, for instance, calls the "riffle" with "easily avoidable" rocks three miles below the Paria a "sloppy rapid, with some bad rocks at head on right." He adds the detail about the Navajo Bridge "almost surely being a record-breaker for height."

12. Freeman, in his original diary, noted that Birdseye operated Kolb's motion-picture camera, which is rather ironic considering Kolb's chronic complaints about photography and Birdseye's broken promises.

LA RUE: Ran all boats through Badger Creek Rapid, except the canvas boat *Mojave*. This boat was carried to the lower end of the rapid where Dodge with life line waited for an upset so he could save the life of a boatman.

BLAKE, AUTOBIOGRAPHY:[13] As I was talking with Moore after the run, he remarked that—"Blake, if you tell me you can take a boat up the side of that cliff, I will believe you. It would seem no more difficult than running that rapid." Gross exaggerating, of course, but I got the message.

FREEMAN, REPRISE: About two miles below Badger Creek we came upon a striking monument of dolomite in the middle of the river [Ten Mile Rock]. A little above this Moore stopped to study what has struck him as an especially interesting section, but made his examination brief after coming upon an outcropping of dead cow—the second we have found in two days. Opening up a long, straight stretch of canyon running east-and-west, La Rue stopped to photograph a great rectangular cliff that towered against the skyline, revealing breaks at either end, neither of which was positively identifiable as the course of the river. [This view, photographed by others, is one of the classic images of Marble Canyon.]

KOLB, REPRISE: After lunch Mr. B. told me to not go more than 3 miles as it was then 2:30 P.M. and find camp. Also to patch Leigh's oar, which I did with Leigh's aid. Broken off at the blade. Soap Creek proved to be less than 4 miles from Badger. We find a good spring from recent rain apparently. Leigh and Freeman are burned like Indians. Blake is nursing a burned shoulder. Clouds are beautifully tinted by the setting sun. After supper we listen in on the radio. Uncle John of the *Times* says he has a sad duty to perform. Mr. Birdseye thought he said something concerning President Harding's death.[14] He read the President's speech prepared for the Night Templars at Hollywood Bowl. Then at 9:15 or 8:15 L.A. time we clearly hear that the President died at 7:30 and for his memory they would go off the air. Later they came on again with a programme in order to give the news of the President's death and though some one posing as the Governor insisted that they go off the air Uncle John stated that he wished to give the news to every one and especially those of the Colorado River party. We sleep with the roar of the Soap Creek Rapid in our ears.

---

13. Blake, "As I Remember," 161.
14. Warren Gamaliel Harding was the twenty-ninth President of the United States (1921–23).

LA RUE, REPRISE: This sad news and the roar of a bad rapid below us made the camp rather gloomy. No songs were sung and we all retired early.

August 3, 1923 — 22,600 ft³/s

KOLB: After our breakfast of oatmeal, pancakes, scrambled powdered egg ham & coffee, I have Dodge cross me in the canvas boat for inspection of the rapid. For safeties sake I decide a short line and portage of both dunnage and boats over the first drop on the west side, then a run for the remainder. Dodge is at the end in each case for emergency, in the canvas boat. Freeman's boat comes next. He doesn't adhere closely to my instructions and nearly hits the rocks. Mr. B. takes a picture with my camera but it balls up so I get nothing of Soap Creek. Took a couple 3As.[15]

FREEMAN: We loaded the boats in expectation of running the rapid, but Kolb returned from across the river to announce that he considered it too dangerous—that a boat would be almost certain to upset in one of the first two big waves, and after that to drive upon rocks not visible from left bank. Everyone was keenly disappointed in missing the chance to make what might have been the first successful run of this notorious rapid.[16] It would not, however, be warrantable to take undue chances with an outfit like ours, especially so early in the voyage. The fall of the rapid proved to be 18 feet. Dellenbaugh and the Kolbs had estimated it at 25. The tendency of the earlier voyagers to over-estimate the height of the rapids, as well as to overrate the boating difficulties, is becoming evident.[17]

BIRDSEYE: It was man killing work to portage the heavy wooden boats [through Soap Creek]. We probably could have run it safely, for ran worse rapids later on, but did not think it wise to take any unnecessary chances so early on the trip.

BLAKE: The Soap Creek Rapid being neatly negotiated, it was

---

15. Kolb owned at least three different cameras that were model number 3A, including a 3A Folding Autographic Kodak Model C Anastigmat, a 3A Folding Special Kodak, and a 3A Graflex. Barbara Valvo, e-mail message to author, March 9, 2006. More information on the Kolb cameras may be found in Christopher C. Everett, "A Historical Study of the Emery Kolb Grand Canyon Camera Collection" (master's thesis, Northern Arizona University, 1981).
16. Soap Creek Rapid was first run by Parley Galloway, with Clyde Eddy as passenger, in 1927. They did not know where they were and surely would have portaged had they known it was the dreaded Soap Creek Rapid.
17. The 1923 USGS expedition would finally determine what the drops were through rapids in Grand Canyon. As Freeman notes here, many previous river runners had greatly overestimated the fall through the rapids.

decided that the best plan of procedure would be to nose[18] the boats partly through and portage the loads for a short distance. My boat was taken through first and was slid for about thirty feet over logs which we had laid over the rocks after it had been let down to a still pocket of water at the brink of the rapid. I took the oars and proceeding through a narrow channel between two boulders I was soon in the swift current. I struck some big waves and let the prow of the boat swing down stream, then I pulled into a big eddy at the foot of the steepest drop. The *Grand* was then let through, following the same procedure, but missed the largest waves. These two boats were through the worst water by shortly after noon, so we ate lunch on the sandy beach where I had landed. After noon the other boats came through, but avoided the roughwater on mid stream. During the afternoon the *Glen* pulled the stakes to which she was moored and drifted out into the eddy while Mr. Word was sitting peacefully in the cockpit writing. When he discovered the movement of the boat he easily pulled to shore. We made camp for the night when we had eaten lunch as there was plenty of wood and clear water. Dodge and Burchard crossed the river in the small canvas boat which we have with us. It rides the waves beautifully since we strengthened the framework and placed inflated rubber tubes on each side just above the water line. We received some word over the radio and "Uncle John" read Freeman's letter in which he said that I had special instructions not to lose the cook.

BIRDSEYE, REPRISE: Set up the radio and heard KHJ clear and well. Received more details regarding the President's death and the funeral arrangements. Heard that August 10 was set aside as a memorial day so decided to lie idle on that day.

August 4, 1923 — 21,700 ft³/s

FREEMAN: We pushed off at 8. There was swift water to the bend, where there began a winding stretch of canyon filled with deep, surging water. The swirls and eddies were of great strength, and it is probable that it was somewhere near the head of this section that Brown[19] was drowned. The boats rolled heavily, and at times were turned completely round. There were several sharp chutes of hard-running water, but none to warrant the name of rapid so freely used

---

18. "Nosing" was a technique Blake learned from Bert Loper.
19. Frank Brown, president of the Denver, Colorado Canyon and Pacific Railroad company, ironically had refused to purchase life jackets for his expedition. His death by drowning could probably have been prevented had he been wearing one. Webb, *Grand Canyon*, 4–5, 9.

by Powell and his successors. In this latter connection, Birdseye has decided to rate as a rapid nothing with a descent of less than six feet in a comparatively short distance. Exception will be made in the case of places dangerous to navigate irrespective of fall.

BIRDSEYE: At noon reached Rapids #3, the first one we could not climb around. The limestone walls rose in sheer cliffs from the water. We did not even have a good chance to look the rapids over and plan how to run them. We named this rapids "Sheer Wall" rapids which has an 8 foot fall and is at mile 13.9. Ran it safely although all got a good wetting. We run the rapids stern first so the boatman is facing down stream. Burchard and I ride on the *Boulder*, Burchard on the rear hatch and I on the front. We all put on life jackets and lie face down, clinging to the life lines. Late in the afternoon ran #4 Rapids which we called 17 Mile Rapids or House Rock Canyon Rapids for it is at the mouth of a canyon from the west which we think is the outlet of House Rock Wash. This rapid is 16.5 miles below the mouth of the Paria River and has a fall of 9 feet.[20] All accounts of previous trips give the distance and fall of the rapids greater than they really are. Camped on the right bank below the rapids. All were too tired to work with the radio so no attempt was made to set up the instrument.

LINT: Sixteen-mile Rapid [House Rock Rapid] has a fall of 8 feet but is nothing but a straight chute of large waves which about half filled the cockpits of three of the boats. I took Moore through it in my boat as he said that he wanted a kick out of it, and he got what he wanted.

BLAKE: During the afternoon, the *Marble*, *Glen* and *Grand* with their usual passengers, drifted ahead of the surveyors and sought the shade offered by large caves into which the boats could be rowed. The air was often filled with song which echoed from wall to wall. The only song omitted from the program being La Rue's favorite, "Barney Google."[21] About 3:30 another rapid was heard ahead and by the sound was known to be a big one [House Rock Rapid]. It was thought the big waves could be missed by keeping near the west shore, but all but the canvas boat were drawn into the main waves. The *Grand* and *Boulder* went so far down stream that they missed the big back flow or eddy and had to be pulled up stream

---

20. The drop through House Rock Rapid changed more in the twentieth century than any other Grand Canyon rapid; Magirl, Webb, and Griffiths, "Changes in the water surface profile."

21. Billy Rose and Con Conrad wrote the "Barney Google" song, which was based on a popular comic strip by Billy DeBeck.

by the painters, to the sandy beach where camp had been made, owing to the fact that there was bed room and plenty of wood, the only place having these desirable qualities which had been seen during the day. Emery dug a well but the river washed into it and spoiled it. I slept under an overhanging rock which formed a roof twenty by forty feet.

August 5, 1923 — 20,600 ft³/s

LA RUE: Got an early start, water quiet for some distance. At Mile 18.1 we came to what we have named Boulder Narrows. A big rock from the canyon walls has fallen in the center of the river. The rock is about 60 feet wide and 50 feet high. The river is cut off except about 60 feet. Nice boating below Boulder Narrows. Smooth water usually means a rapid ahead. No exception in this case. We soon heard the roar of a rapid [North Canyon Rapid]. E. Kolb looked this one over from both banks and then studied the layout of rocks both above and below. The first decision was that we should have lunch. We lunched in the shade at the mouth of a side canyon on the L.B. A few stories and a little excitement caused by some of the bunch trying to smoke a chuckawalla out of a crack in the wall. After Emery had eaten a good lunch he decided to run Rapid No. 5 (Twenty Mile Rapid) [North Canyon Rapid], Fall 11 feet. Took stills and movies.

FREEMAN: Immediately on landing to look it [North Canyon Rapid] over, Kolb said he recognized it as the one at which he and his brother had portaged, and which he had mistakenly thought was the one we ran last evening. It appeared scarcely less savage than Soap Creek in its first sharp drop, while below, around a bend, was the head of a second rapid which seemed to be jumping at a very lively rate. Kolb looked over the rapid both sides before deciding to run. He favored a course through some shallowly covered boulders along the right side rather than down the clearly defined V, which he reckoned would take a boat into the heavy waves against the left wall. With Lint on the stern of the *Marble*, he ran it this way, and without getting into water as rough as he had encountered last evening. A slightly submerged midstream boulder near the lower end of the rapid was narrowly missed, however, before he pulled over to the left bank. A heavy sand storm was blowing out of the side canyon as I drifted down to the head of the rapid, and in pulling against it I got too far to the right and found the *Grand* drifting down on to a vile mess of boulders almost against the bank. Hard pulling did not quite carry me back to the channel, and as a consequence a corner of the stern bumped solidly against a rock, allowing the boat to swing completely around. The *Grand* drove through the next couple of waves and then passed to the side of the big rollers against the wall under good control. Pulling through the current of the tail of the rapid, I beached the boat on the left bank, to find that she had suffered no damage from the collision.[22] The *Grand* probably had the driest run of any of the boats; nevertheless, getting in where I did at the head was not creditable, even in the face of the sand squall.[23]

KOLB: Dodge was stationed on the south shore [of North Canyon Rapid]. We hit the breakwater on the west dropping close to the big rocks to avoid the mess of waves. I make it perfectly. Blake follows makes a clean run. Then Lint. I make movie of both. Freeman does not get the proper channel. Blake had joined me. When I saw where Freeman had started, my pulses leaped. I told Blake the *Grand* would smash to pieces. Freeman after dropping over the 1st rock saw his predicament and used every ounce of strength to pull out. He did well, but dropped and struck a huge boulder whirling him around but afterwards proved to have done no damage. A half mile below we run another V rapid. Burchard on Leigh's boat gets soaked.

LA RUE, REPRISE: About half a mile below 20 Mile Rapid we came to Rapid No. 6. Fall 5 feet. We all rode the decks. Got wet but no further damage. Camped at 21.5 miles below Paria River. Set up radio. Aerial suspended from canyon wall by cord about 300 feet long. 200 feet of wire used as the aerial. Got program of May McDonald Hope who played the piano for—Louvaviski, violinist;[24] "Uncle John's" words regarding the funeral train carrying the body of Pres. Harding through Wyo. Train delayed 15 minutes due to broken rail.

August 6, 1923 — 20,100 ft^3/s

FREEMAN: This has been from a boating standpoint our liveliest and most eventful day. Six or seven major rapids and a number of riffles have been run, in nearly every instance the head of one being in sight from the foot of the last. The total distance run was three miles, with a descent of 50 feet. This may well be the heaviest average

---

22. The *Grand*, the only new wooden boat, was far sturdier than the others and suffered impact with far less damage. Kolb likely assigned the boat to Freeman knowing that, as the least skilled boatman, he was the most likely to hit rocks.
23. In Freeman's original diary, this sentence reads, "I am inclined to think I would have done better with my heavy boat and load by running the V, nevertheless my getting in where I did was not creditable, even in the face of a squall." Freeman diary, p. 48.
24. The correct name for the violinist is Calmon Luboviski. The concert was reported in "Artists Excel on KHJ Sunday," *Los Angeles Times*, August 6, 1923 (accessed February 27, 2006).

fall in Marble Canyon.[25] As the *Grand* approached the second rapid, with Moore and La Rue riding, Kolb, who had landed to look it over, signaled for me to go on through without stopping. It was a straight run down the middle with no chance of rocks—only big water. Kolb's direction was to pull down the V, and then to the left [today, this is known as a Major Powell move]. This was quite correct, only, knowing there was no real menace, I did not pull to the left hard enough. The result was that side-running waves of astonishing weight and solidity came rolling over the *Grand*, completely filling the cockpit. She still rode fairly steadily, but so low that all of the succeeding waves swept her fore-and-aft. She was still readily manageable with the oars, however, and I had no trouble in bringing her to the left bank in an eddy well above the head of the next rapid. Ten minutes baling had the cockpit dry again. The water-tight box under the seat in which our cameras were carried proved as good as its name. My small camera, which was not in the box, was thoroughly soaked, however.

LA RUE: In Rapid #9 Dodge in the canvas boat turned over; for a moment both boat and boatman were under. When they appeared the boat was bottom up and Dodge hanging to the side. He turned the boat over and climbed in. The boat did float although full of water, thanks to auto tire tubes on the sides and oil cans tied in the boat. Dodge finally got one oar loose and paddled into an eddy. Two other boats were sent to his rescue. However, he saved himself.

BIRDSEYE: Dodge is the strongest swimmer in the party and afraid of nothing wet. We camped on the right bank near some good springs at the head of #13 Rapids. We named the rapids at this point "Spring Cave Rapids."

LA RUE, REPRISE: Our camp tonight ... is just above the cave ... where the Kolb Bros. spent two days in Nov. 1911. Some one has been this way since 1911 for we found in this cave the articles listed below: 1 Rubber boot, 1 Bucket, 1 Saw, 1 Brace, 3 Bits, 1 Coffee Grinder, 1 Shovel, 15 Steel traps, 1 Hammer, 1 Pair tin shears, 1 Stew pan, 1 Gig, 1 Rivet punch, 1 axe, 1 awl, 1 plate, 1 flour sieve, 1 piece leather, nails. Took pictures of the cave.[26]

---

25. Freeman is correct; the closely spaced rapids, known as the Roaring Twenties, represent the steepest fall through Marble Canyon.
26. Following the publication of his "Surveying the Grand Canyon of the Colorado," Freeman received a letter from Fred T. Barry of New York regarding the collection of materials found in the cave at Cave Springs Rapid. Mr. Barry wrote, "It may interest you to know that I was one of a party of three persons who cached this outfit where it was found, during the spring of 1888 and we got out of the canyon

R. C. Moore sketch of Marble Canyon from above Rapid 17, now the unnamed rapid upstream from the Fence fault.

R. C. Moore diary, p. 23, courtesy of the University Archives, Spencer Research Library, University of Kansas Libraries.

August 7, 1923 — 20,100 ft³/s

DODGE:[27] Freeman had advocated the use of a 14-foot folding canvas boat for the rodman so I started off in that with my tongue in my cheek. It was easy to portage around rapids such as Badger and Soap Creek but below that I tried to run it through a secondary rapid [Cave Springs Rapid] and the first wave slapped me out of it, filled the boat, and we went through side by each. The excitement that little upset raised makes me smile now. However, it wasn't long until, in lining it down a rapid difficult to portage, someone either held a line too long or let go a line too soon. Anyway, it was crushed in the rocks and all we salvaged were the oarlocks.[28]

---

without any casualties. It was our intention to trap the following winter, but never did so;" Fred T. Barry to Lewis Freeman, June 6, 1924, box 70, folder 1, Marston Collection.

27. Dodge, *Saga*, 36.
28. Kolb, in his diary, indicated Burchard was the one who failed in his timing on the line. In the *Grand Canyon Film Show*, Kolb stated "One of our engineers attempted to line a boat around this point without my notice, and this is what occurred to him, smashed to pieces. We saved no part of it."

BIRDSEYE: We surely hated to lose the little boat for it was a great convenience to have it for the rodman and it always gave us a feeling of safety to have it at the foot of a rapids ready to go to the rescue of any boat that needed help. We decided to order another by wire from Hance Trail to be delivered at Bright Angel but later changed our minds when we got into the heavier water in the Grand Canyon.

FREEMAN: All four boats ran the rapid [Cave Springs Rapid] without passengers. The waves were rough, but after starting down the middle of the V it was not hard to work to the right into a big eddy filled with an almost solid mass of driftwood. Blake took on Dodge and his rod, while the cook transferred to Kolb. Rapids continued as frequently as yesterday, but there was none that was especially threatening. Running a sharp pitch just before lunch, I managed to avoid some of the heaviest waves by hard pulling, but in working out of an eddy setting against the righthand cliff below I had to lean on to my oars rather heavily. Nothing gave way at the time, but a few minutes later my right oarlock broke off at the ring, where there proved to have been the flimsiest sort of a weld, just like that of those previously broken by Blake and Kolb. As a faulty oarlock usually gives way in a place where loss of control may well mean the loss of a boat, this utterly worthless smithy work is a rather serious reflection on Fellows and Stewart who provided them. There was a slightly greater interval between rapids this afternoon, and longer spells of quiet water. We ran everything with passengers. The last rapid above camp followed the bend around under the righthand cliff, with some threat in a projecting point of rocks near the lower end. All the boats ran into the V bow-first, then swung and pulled away from the main line of waves and the rocks below. The *Grand's* passage was dryer than usual. The splash did not even put out La Rue's pipe, which has come to be accepted as the standard test of a sloppy run. We have made a very pleasant camp on a beach between two winding rapids, with the seep from a canyon supplying good drinking water again [this camp appears to be at the Fence Fault, about mile 30]. Burchard is inclined to believe that we are in the vicinity of Stanton's "Point Retreat," and that the canyon behind our camp is the one by which he left the river after the culminating disaster had drowned two more of the Brown party of 1889.[29] Distances and data are too conflicting and meager to make a conclusion possible on this point at the present.

---

29. Burchard was correct; the Brown-Stanton expedition, minus Brown and two others, evacuated here in July 1889; Smith and Crampton, *The Colorado River Survey*, 86–90.

The wreck of the *Mojave* at Cave Springs Rapid, August 7. Burchard and Kolb are looking at the wreckage of the skin; Dodge is looking despondent at right.

E. C. La Rue 57, courtesy of the U.S. Geological Survey Photographic Library.

BIRDSEYE, REPRISE: Ran four other rapids numbered from 14 to 17 with falls ranging from 5 to 7 feet [these would include Tiger Wash and 29 Mile Rapid], and also made two dam site surveys at sites Nos. 1 and 2. Worked on the survey of Dam Site #2 until dark and finished it the next morning.[30] Were too tired to put the radio.

KOLB: La Rue asks for a survey for dam site. B is peeved. They get to camp just at dark.

BLAKE: Tonight is the first camp we have made where no rapid roars to remind us of the power and danger of the Colorado.

August 8, 1923 — 19,500 ft³/s

BIRDSEYE: Passed Vasey's [Vasey's Paradise] on the right bank at mile 31.1. This is described by Powell-Dellenbaugh and Kolb. A large stream of clear cold water gushes from the cliff about 75 feet above the river level and is surrounded by green brush—some of it poison oak [ivy] We landed above and climbed under the stream to

---

30. La Rue named this the Redwall Dam Site, and a description appears in La Rue, *Water Power and Flood Control*, 53–56.

La Rue with his cameras at Vasey's Paradise in Marble Canyon on August 8. La Rue, the official USGS photographer for the expedition, took both panoramic and motion picture images at Vasey's; in this view he appears to be using the former, likely an Al-Vista model.

Lewis Freeman photograph, Topography D 27, courtesy of the U.S. Geological Survey Photographic Library.

fill canteens. About 500 feet above Vasey's Paradise and on the same side of the river is a large cave 80 feet above the water and 150 feet deep [Stanton's Cave]. Explored this and found no evidence of it ever having been visited before.[31] Surveyed #3 Dam site at mile 31.7 [above Redwall Cavern on the right].

FREEMAN: La Rue asked me to land him on the boulder island for a panorama of the falls [Vasey's Paradise] and this was managed by running down the right side of it and pulling back up the eddy at the lower end. We took several pictures, including a movie with the *Boulder* passing. Kolb semaphored us to look for a brass tripod he had left on the same bar twelve years ago, but we found no trace of it—hardly surprising considering the many floods that have passed over it since. We stopped to explore a large undercut cave or amphitheater beneath the left cliff that was easily large enough to have seated from 10,000 to 15,000 people for a concert [Redwall Cavern].[32] The roof rose steadily until it overhung the middle of the river at a height of 200 feet. The floor was a succession of rising terraces of smooth, hard sand, rising like the seats of a stadium until the highest touched the vault of the limestone roof. We took photos, both from the outside and inside, and La Rue made a movie.

BLAKE: We had shade until 11 o'clock this morning due to the high walls of the canyon.... the afternoon's journey continued thru one of the most picturesque canyons imaginable. Many caves and springs of sparkling clear water were passed. Once as we were passing between two sheer walls of limestone we heard a sound as of excaping [*sic*] gas or air and seeing some mint or similar plant went to investigate and found a fountain of water which had enclosed itself within a wall of lime which extended from about ten feet up on the face of the cliff to the water's edge, the only opening to the outer air being a jagged hole in the enclosing crust near the surface of the river [this spring still is an attraction to river trips]. This hole enabled us to get to hand hold and hold the boats still while canteens were

---

31. Stanton used the cave downstream from South Canyon to cache supplies before retreating from his ill-fated first expedition. Split twig figurines were located here in 1934; Robert C. Euler, *The Archaeology, Geology and Paleobiology of Stanton's Cave, Grand Canyon National Park, Arizona* (Grand Canyon, AZ: Grand Canyon Natural History Association Monograph No. 6, 1984).

32. In his original diary (p. 61), Freeman estimated that 20,000 people would fit in Redwall Cavern. He estimated far fewer than John Wesley Powell, who stated that 50,000 people could fit in the cavern; John Wesley Powell, *Canyons of the Colorado* (Meadville, PA: Flood & Vincent, 1895); facsimile edition published as *The Exploration of the Colorado River and Its Canyons* (New York: Dover Publications, 1961), 237. Citations are to the Dover edition.

filled with the cold water. Another place of interest was a cave washed by the river, which extended nearly two hundred feet into the wave of the cliff, and which, during high water, would make voyaging by boat very dangerous. Just before making camp the most striking natural castle ever seen, was passed. Towers, windows and balconies had been formed by the elements and a natural bridge spanned a gulch which separated the castle from the main cliff [Bridge of Sighs, first noted by the Kolb brothers in 1911]. It was suggested that this weird looking wonder of nature be called "Goblin's Castle," which seemed most appropriate. Camp was made just below and opposite the "Goblin's Castle" and at the head of a rapid [36 Mile Rapid]. Our beds were laid out on the solid limestone which rises in tiers of flat benched, making many single and some double sleeping apartments.

August 9, 1923 — 18,100 ft³/s

BIRDSEYE: Passed some beautiful arches in the right wall at mile 41.2. We named these the Royal Arches. Camped on a sand bar on the left bank at the head of #19 Rapids which we named "Boulder Rapids" on account of a large boulder as big as a small house in the center of the channel.[33] The entire force of the river seemed to drive towards this boulder and it looked bad.

FREEMAN: Moore found a number of fossils (or rather hitherto undiscovered ones) in the limestone near last night's camp; also that the Muav is beginning to appear under the Redwall. Late in the forenoon he reported the first appearance of the Bright Angel Shale. About 11 this morning we landed to discover deer tracks, buck, doe and fawn. At the same point the tracks appeared we found a number of Judas Trees [catclaw acacia?];[34] also the first mesquite. We lunched under the cliff where the deer were sighted and continued the survey from there in the afternoon. Two large canyons [Buckfarm and Bert's] coming in from the right inside of a mile took some time to survey. As we had planned to halt over tomorrow—the day of the Harding funeral—some care was taken to pick as desirable a camp as possible. Two or three large alcoves, with springs and verdure, were reached too early in the afternoon to make a stop practicable.

---

33. The name Boulder Rapid was originally applied by Robert Brewster Stanton in 1890; Smith and Crampton, *The Colorado River Survey*, 137–38; Webb, *Grand Canyon*, 177. The 1923 USGS expedition also called it Boulder Rapid at first, despite Kolb's misconception that the boulder appeared after his 1911 trip. After the expedition, Birdseye renamed this drop President Harding Rapid.
34. In his original diary, Freeman noted that the "Judas Trees" were "like lilacs, with brown pods," a description that could match *Acacia greggii*, the catclaw tree.

We finally halted about 5 where a rapid, blocked squarely by a huge fragment of limestone from the upper wall, presented an obstacle that could not be passed without looking over. The fracture on the righthand cliff from which this, with several other great slabs of limestone have fallen, is plainly very recent.[35] Kolb says he does not remember having seen it in his voyage of 1911,[36] and the force of a river expending itself in an up-spouting geyser against a mid-stream boulder is not a thing readily to be forgotten. As the channel is much narrowed by a jutting bar of boulders from the left, the blocked chute will have to have careful study before running, if it is runnable at all. The cliff from which the obstructing rock has fallen is one of two tentatively marked as "Point Hansbrough" on our special map. Stanton gave that name to a jutting point at the foot of which he buried a man of that name, lost about ten miles above on the Brown voyage. The remains were found and identified shortly after Stanton resumed the interrupted voyage in the winter of 1889–90. We have not the data to identify the point beyond doubt and so will not make an exhaustive search for the Hansbrough inscription.[37]

LA RUE: While landing to give a rod reading I saw deer tracks on the left bank. They could not go upstream without swimming so we watch the narrow talus forming the bank on the left. About a mile below we saw the doe and fawn.[38] The doe ran back up the rim so we set up our movie & still picture cameras and signaled to the other men to chase the deer back. No luck. When the deer found she was between the men on shore she took to the river and swam to the other side. Leigh Lint tried to row a boat and catch her in the water. However, the deer could swim too fast to be caught that

---

35. Freeman thus incorrectly established the origin of the rock in President Harding Rapid as coming from a rockfall from river right. Indeed, in the late 1990s, rockfall closed off the right run, which the USGS expedition took advantage of in 1923, but those rocks expanded the bank and didn't roll to the center of the channel. It is more likely that the boulder was rafted into the river during a debris flow from the relatively large canyon on river left; Webb, *Grand Canyon*, 177. In all likelihood, we will never know where that specific boulder came from, but the constriction that forms the rapid is maintained by debris flows from river left.
36. Kolb was mistaken. The rock was definitely present in President Harding Rapid in 1911; Stanton had photographed it in 1890. Webb, *Grand Canyon*, 177.
37. Peter Hansbrough's grave and inscription are on river left below President Harding Rapid and at the base of the Muav Limestone cliff above the commonly used camp. The 1923 USGS expedition camped about a third of a mile upstream from the gravesite.
38. Deer remain a frequent sight along the river corridor in this reach. In 1923, the deer herd on the Kaibab Plateau—west of the river corridor at this point—explosively increased, leading to the need to cull the herd in the 1930s; John P. Russo, *The Kaibab North Deer Herd: Its History, Problems and Management* (Phoenix, AZ: State of Arizona, Game and Fish Department, Wildlife Bulletin No. 7, 1964).

Surveying at 36 Mile Rapid. Burchard is probably the man at the
alidade, with Birdseye recording. Freeman snapped this
striking image about August 8.

*Lewis Freeman photograph, Topography D 32, courtesy of the U.S.
Geological Survey Photographic Library.*

way. We could have shot her but did not have the heart to kill a doe with her fawn. The buck beat it. We didn't even catch a sight of him. We had visions of a big feed for we planned to lay off the 10th to observe the passing of our late President.

BLAKE: Camp was made on a sandbar near a narrow portion of the river, where the current becomes swift and dashes against a huge boulder in mid stream. The wind blew pretty hard before supper and the sky is overcast with dark clouds in the west. The "radio bugs" are stringing the aerial so as to amuse themselves during the evening. Just before dark a heavy rainstorm struck camp and everyone tried to find shelter. Freeman stripped and put his clothes under cover, Emery got into the front hatch and pulled the cover over the hatch. Leigh and I stood in the front hatch of my boat with a tarpaulin over our shoulders, but the cold rain running down our backs caused me to crawl into the hatch and pulled the tarpaulin over us. We went to sleep for awhile, but as the wind ceased its fury Emery came forth from his retreat and woke us up to see whether we had smothered or not. La Rue had become so cold that he had started the fire and was hovering over it, though he wore only a shirt and held some kind of cloth or a piece of canvas over his head. Freemen had dressed up somewhat by donning a life preserver which was his only article of apparel. After getting warm by the fire everyone rolled his bed out and did his best to keep dry.

August 10, 1923 — 17,200 ft³/s

KOLB: We get up late. Eat about 9 am. What a fine breakfast. Oatmeal, hot cakes, bacon, syrup & coffee. We bail rain water from our boats. The boys play ball while I cross to size up the rapid. It doesn't look good. The men wash their clothes, write letters & notes. I make comic movies. Dodge trims my whiskers.

BLAKE: We are resting today, in respect to President Harding. Leigh and I went up a side canyon this morning and filled the canteens and water bags with clear water which was falling from the rim a hundred feet or so above. We also washed our clothes. I mended shoes. Freeman, Moore, Dodge and Lint played tennis, using a rubber ball. They had no net and used their hands for rackets. La Rue and Emery each took some moving pictures. Dodge set a big drift of wood on fire.[39] Emery, Leigh and I went swimming.

---

39. This is one of the first documented examples of burning the enormous piles of driftwood that used to be found in Grand Canyon. Some years after the expedition, pyromaniac river runners began to call themselves Drift-Wood Burners (DWBs). In order to become a "member" of this "club," one had to ignite a pile with a single match;

August 11, 1923 — 17,000 ft³/s

BLAKE: The instrument men, rodman, and cook went ahead this morning while the boatmen crossed the river to inspect the channel between the big rock and the west shore [at President Harding Rapid]. Emery then came thru, barely striking the edge of the back wave thrown out by the rock. Leigh then came thru, going somewhat closer to the rock. I came through next, and as the boat struck the big back swirl of water it turned on its side throwing me clear out of the boat and at the same time snapping an oar lock in two. A couple of strokes put me in reach of the boat and in a few seconds I was over the side and had a spare oar in place. The lost one was picked up in an eddy below. Another doe and fawn were seen today.

LINT: [Blake] turned a complete flip flop and landed feet first in the water but he wasn't long getting hold of the gunwale and climbing aboard. Freeman hit the same place that Blake did but his boat was so heavy that it didn't affect it.

BIRDSEYE: Had fairly quiet water during the rest of the day, running only several riffles too small to be dignified by the name of rapids. Passed 3 large caves on the right bank at mile 46.5 which were beautifully arched and pillared. We called these caves "Triple Alcoves." Passed Saddle Canyon coming in from the right at mile 46.9. This was filled by a rock slide up to the 3200 foot level so we did not have to traverse up it far. Camped at the mouth of Little Nankoweap Creek coming in from the right at mile 51.6, about on the north edge of the Grand Canyon National Park. Set up our radio with aerial at right angles to the direction of Los Angeles. Did not get KHJ for the first time but did get the Deseret News of Salt Lake City (KZN) better than we had since leaving Lee's Ferry.

LA RUE: "Barney Google" came in strong.

August 12, 1923 — 20,600 ft³/s

LA RUE: Necessary to carry survey up Nankoweap and Little Nankoweap Creeks. This gave some of us time to scout around a little. Emery Kolb, Blake, and I went up to some Cliff House, on the north wall 700 feet above the Colorado R. [the Nankoweap

---

Robert H. Webb, Theodore S. Melis, and Richard A. Valdez, *Observations of Environmental Change in Grand Canyon, Arizona* (Tucson, AZ: U.S. Geological Survey Water Resources Investigations Report 02-4080, 2002). In 1940, Barry Goldwater, later to become U.S. Senator and presidential candidate, was one of the most avid DWB practitioners, thanks in part to the influence of pioneering river guide Norman Nevills.

granaries]. Took 3A and stereo pictures. I measured the flow of Nankoweap Creek about 3 sec. ft. [ft³/s]. This is the first stream to enter the Colorado below the Paria. No rapids today. Water very swift with many riffles; some of them bad. At the mouth of Nankoweap I picked up a Log Cabin Syrup Can which I threw in the river at Lee's Ferry July 24th. I had scratched my name on the can. It was on a sandy beach about four feet above the present water surface.

FREEMAN: We camped on a fine hard sand-bar above the dry bed of the Kwagunt. The latter has thrown out an extensive boulder fan, forming a rapid which (from the right side at least) looks as though it would need very careful navigation. The radio was set up with one end in a high mesquite tree, and carefully orientated toward Los Angeles. KHJ came in strong, probably the best we have ever had it in spite of the fact that the walls are now over 4000 feet above the river. Moore gave a complimentary concert on his own account by rendering, with La Rue, a parody of his own composition on "Gallagher and Shean."[40] I regret to state that a comparatively small portion of it was fit for inclusion in this chaste record.

KOLB: Before supper we get chilled. Soon after the temperature rises again but night is fine for sleeping. A big rapid is roaring below us. Cape Solitude is in view. This is a beautiful camp.

August 13, 1923 — 24,700 ft³/s

BIRDSEYE: Traversed up Kwagunt Creek which flowed clear water down to about ½ mile of its mouth. During this time the boatmen ran #21 Rapids (Kwagunt) which has a fall of 6 feet. Ran #22 rapids with a fall of 5 feet, at mile 58.8 [60 Mile Rapid]. At 3 P.M. we reached the mouth of the Little Colorado River at mile 60.5 below the mouth of the Paria River and found it in flood, too high to measure. We found the water level of the river at this point to be 2725 [ft] which was about 20 feet lower than indicated by interpolation from the Vishnu [15'] sheet and from vertical angle readings by the Southern California Edison Company. (After adjusting our line at the B.M. [benchmark] at Hance Trail—15 miles below—where we checked 8 feet high we made the adjusted elevation of the water at the mouth of the Little Colorado to be 2718). We set a bench mark—cross cut in rock above high water and marked by a cotton wood pole in a rock mound.[41] Mileage and numbers of rapids from

---

40. Ed Gallagher and Al Shean (Albert Schoenberg) were a popular vaudeville team that performed on Broadway and in the Ziegfeld Follies of 1922.
41. Unfortunately, this benchmark has never been found. Its location would greatly help long-term monitoring efforts for the longitudinal profile through Grand Canyon.

here on begin at the mouth of the Little Colorado,[42] where Marble Canyon ends and Grand Canyon begins.

FREEMAN: The rapid over Kwagunt bar appeared almost as rough from one side as the other, but the most favorable course seemed to be down the V near the lefthand bank. Kolb, Lint, myself and Blake ran in the order named. All had a bumpy passage but none was in any difficulty. We drifted in on the left side of the V in order to miss two big corkscrew waves that looked equal to upsetting a boat. This still left a jumbled line of combers that tossed the boats about merrily and doused a lot of water abroad. With the current setting sharply against the left bank, we had to pull away hard to make the eddy on the other side. It is a rapid that would brook no carelessness. It is evidently much rougher than the Kolbs or their predecessors found it—doubtless the result of a considerable augmentation of the boulder bar during a torrential flood from the Kwagunt.[43] While Birdseye and Burchard remained to run a line up the Kwagunt, I pushed off with the *Grand* for a direct run to the Little Colorado, which La Rue wanted to measure if possible. Riffles were numerous—some rough enough to throw water over the boat. We were an hour on the run. Point [Cape] Solitude loomed higher and higher as we drew near, and when we ran down under it at 11 the somber crest of the great headland towered half way to the zenith. I land La Rue on the right bank for a panorama looking into the mouth of the Little Colorado. Moore accompanied him to geologize, while I stayed with the boat and took a cooling dip—probably the last for many days, as the ill-smelling Little Colorado appears to be pouring out one of its nauseous floods. Lunching under a shaded shelf, we waited until the *Marble* came along and then pushed off, ran a shallow, winding riffle and pulled into the quiet water of the lesser and lower of the two mouths of the Little Colorado. It was flowing an unfragrant gray-green liquid mud which La Rue estimated at about 1500 second-feet. As the marks on the banks showed it had been five feet higher within a day or two,

---

42. In their diaries, members of the Birdseye expedition, and particularly Claude Birdseye, give mileages from specific fixed points defined by where benchmarks had been previously established (e.g., at the mouth of the Little Colorado River). By the time the plan-and-profile maps were published, the compiler (likely Burchard) turned the mileage into continuous distances downstream from Lee's Ferry; see U.S. Geological Survey, Plan and Profile Maps.

43. Unfortunately, La Rue did not photograph Kwagunt Rapid in 1923, nor did any other photographer from 1872 through the early 1950s. Based on Emery Kolb's mistaken memory at President Harding Rapid, it is difficult to believe his interpretation of change at Kwagunt Rapid.

he reckoned it had been up to at least 10,000 second-feet—equal to the normal flow of the main river at this season.[44] ... we pushed on down the river. The latter has been rising steadily for a day or two, and, augmented by the flow from the Little Colorado, ran with a weight and force that was different from anything yet experienced. It is doubtless the consequence of an unusually high silt and sand content in the water—a regular rising stage condition.[45]

BLAKE: I landed on the east side of the river to pick up Moore, and in doing so received a semaphore message from across the river that there was a spring of water below me, and to fill the canteens. I took water bag and other containers and went to the spring which bubbled forth so beautifully as to make me wish to partake of it immediately, so I filled a can with it and took a deep draught only to spit it forth and make a wary face and a startled exclamation. The water was so salty that no one could swallow of it when I filled a pail and offered them a drink.

KOLB: After lunch, I proceed with the cook and camp on a projecting sand & rock bar just past where walls sheer up about 1½ miles below the Little Colorado [Crash Canyon]. A storm is brewing the sand is blowing. Burchard's makeshift tent just went down. The wall is full of holes, shelves & little grottos. La Rue has Freeman picture him in various poses.[46] La Rue is scorching Freeman today on account of laziness. Tells the cook when he reached the Little Colo. he would tell F. how useless and lazy he was. Calls him G.D. flunkey and sites [*sic*] himself as a technical man. I hear the boats bumping about 11 P.M. Wake the men and have the boys pull them up and tie to long line. It rains quite hard. Overhanging cliff helps keep our beds dry.

---

44. The 1923 expedition found the Little Colorado River to be in recessional flow from a flood peak on August 13. As we shall see, this was a harbinger of coming events: the Little Colorado River watershed received high rainfall in the summer of 1923, which led to its flood of record in September and its consequences on the expedition at Lava Falls.
45. On the San Juan River, the authors have experienced firsthand the effect of high sediment concentration on boat mobility. Summer storm runoff, heavily laden with sediment, increases the density of flow, making it more difficult to row boats.
46. Debris flows initiated in Crash Canyon in September 1990, and we spent considerable time in the camp that the 1923 expedition used; Melis, Webb, and Griffiths, *Magnitude and Frequency Data*. Indeed, we found the ledge La Rue posed in for Freeman's photograph.

5

Surveys and Portages

Furnace Flats through the Inner Gorge

On August 13, 1923, the 1923 USGS expedition camped at Crash Canyon, a small tributary on river right at river mile 62.6. This small canyon became infamous in 1956 after two commercial airliners collided over Grand Canyon and the debris spread over this drainage and several smaller ones nearby. Just downstream, the Colorado River corridor opens into a reach fondly known as Furnace Flats for the extreme temperatures encountered in summer. This reach spelled more work for the surveyors, since any dams built downstream would impound reservoirs that would expand to this stretch.

The expedition had to deal with frequent thunderstorms, which mostly occurred in the middle of the night. They had learned to climb into the waterproof hatches instead of setting up tents that blew down during wind gusts. Dodge showed his hardiness, giving his sleeping bag to the cook and sleeping nearly naked under a piece of canvas. Others sought overhanging ledges to sleep under. The expedition members experienced the typical summer phenomenon of being hot during the day and cold during the storms. They also experienced something that does not happen often in Grand Canyon in the twenty-first century—sandstorms.[1] They also proved the experts wrong with their usually clear radio reception of news reports and entertainment programs from a variety of cities in the western United States.

Sand, lodged in everything on the expedition, was a continuing irritant for the crew, who were beginning to fight among themselves.

---

1. Webb, Melis, and Valdez, *Observations of Environmental Change.* Reduction in sediment concentration in the river, created by the mere presence of Glen Canyon Dam, has reduced sediment along the river channel and reduced the potential for blowing sand.

View upstream of Granite Rapid from below the mouth of Monument Creek, mile 93.6.

Tom Brownold photograph, Stake 1462, Desert Laboratory Collection of Repeat Photography.

Both Kolb's and La Rue's movie footage showed what could be interpreted as a harmonious crew, smoking like chimneys—there was no shortage of tobacco on the expedition. But Freeman criticized La Rue, La Rue bad-mouthed Freeman, other crew members called Freeman lazy, and even La Rue's wife had problems with her husband's behavior. Kolb's temper boiled over concerning photographs and movies, and he threatened to quit the expedition. Showing his remarkable leadership skills, Birdseye was able to keep the peace and hold his fractious expedition together.

As noted in his diary, Birdseye chose to begin renumbering both rapids and river miles at the Little Colorado River Confluence, hence Lava Canyon Rapids, which modern river guides place at river mile 65.5, is identified as "Rapids #1 at mile 4.1."

August 14, 1923 — 28,100 ft³/s

FREEMAN: A long series of winding riffles occurred just below camp [Crash Canyon]. In running one of these bow-first Moore, who was busy with geological notes, was all but knocked off the boat by an unexpected wave. I had my usual shower-bath. The *Grand* seems to take rather more than her share of water over the sides into her cockpit. This may be due to her lines, or to the fact that she is unduly low in the water. At any rate, I have not had a dry pair of pants for more than half an hour at a time since we left Badger Creek.

LA RUE: Took cross section today at what might have been a dam site (Dam Site #4) 3 miles below L. Colo. The soft shale in both walls ruined this site. No possibility now until we near Hance Rapid.[2]

KOLB: Distance today 7.4 to Tanner Canyon. On opposite bar I light drift that the people at Desert View might see it.[3] Our river seems different current & boils are worse. Rapid [Lava Canyon Rapid] at McCormick mine[4] has some pretty big waves. Get clear water in pool there. Take a wash there. River water is very muddy, worse than at any time since leaving Lee's Ferry. I develop severe pain over my kidneys. Leigh doctors me up with Sloans liniment. As Frank is cooking supper, a terrific rain comes up. Two striking rainbows arch over Comanche point and last ¾ hr. No rain during night. Burchard did not want to camp here. He, no doubt, is pushing things to hear from his wife who brought him a son just before leaving. We are about a week ahead of our intended schedule. At times they would push on uninspected rapids, jeopardizing the boats if their nervous anxiety had the desired effect on the boatman.

---

2. Hance Rapid, over twelve miles downstream from Dam Site #4, was named in honor of "Captain" John Hance, a colorful Grand Canyon pioneer who earned his living initially in mining, and later, in the more lucrative tourist market; Anderson, *Living at the Edge*, 60–67.
3. In 1923, Desert View was a lovely canyon viewpoint along a bumpy dirt road. The distinctive Desert View Watchtower, visible from the river, was not constructed until 1932, using a design by Fred Harvey's favored architect, Mary Jane Colter. J. Donald Hughes, *In the House of Stone and Light* (Grand Canyon, AZ: Grand Canyon Natural History Association, 1978), 87–88, 97.
4. The McCormick mine was one of several workings belonging to a consortium of miners, including Seth Tanner, for whom the Tanner Trail is named. George H. Billingsley, Earle E. Spamer, and Dove Menkes, *Quest for the Pillar of Gold: The Mines and Miners of the Grand Canyon* (Grand Canyon, AZ: Grand Canyon Association, 1997), 74–77.

LA RUE, REPRISE: Fierce thunder storm over head. Just got my canvas stretched over some logs and am sitting under it with the rain pouring all around me. Would not be so bad if my canvas didn't have so many holes in it. The thunder roars up and down the canyon like 1700 big guns cut loose at once. Pretty rough water today. Really dangerous on account of high waves and whirlpools (Suck holes). There will be bad water tomorrow. All well except Col. Birdseye has a bad toe, infected and swollen double its normal size. The cook, Frank Word, has trouble with his eyes, which makes him feel miserable. The rest of the bunch seems to be O.K. except Emery Kolb who has a touch of lumbago. A bad night for lumbago.

August 15, 1923 — 31,200 ft³/s

LINT: Ran the Tanner Creek Rapid (24) without any trouble although it had large waves and a few rocks in it. 11-foot fall. Picked up an old almost worn out circular sea-style life preserver just after lunch. While pulling in to shore this afternoon I hit a submerged rock and as Mr. Birdseye was not quite balanced it knocked him off in the river but he grabbed a life line and held on until I pulled into shallow water. No damage done except the Colonel getting a good soaking.

FREEMAN: We ran in swiftly descending water all morning, and about noon came to the head of an unfriendly-looking rapid of which Kolb had spoken several times [Unkar Rapid]. He had lost an oar in running it, with some consequent trouble in avoiding the cliff. This rapid describes a complete crescent around a sandy bench, surging hard against the sandstone cliff on the left during all of its upper half. It is continuously rough water for nearly half a mile, with most of the fall at the upper end, where Burchard found the drop to be 21 feet. Several deer were scared up as we skirted the bench in looking over the course. A pile of driftwood near the head, several acres in extent, was lighted on the chance it might be seen from the rim and give warning of the fact that we were approaching the foot of the Hance Trail several days ahead of schedule. We four boatmen ate a light lunch by the boats above the head of the rapid, and were about to cast off when a succession of heavy wind-squalls came scurrying down the river and held us up until they had passed. At the same time what was almost a wave of water came rolling down— probably from some cloudburst above—so that the already rising river surged up more than a foot in a few minutes. With the channel filled with floating driftwood, it became imperative to run at

The crew, boats, and camp kitchen at a camp at 75 Mile Canyon above Nevills Rapid, on August 15. From left to right: Lint, Blake, unidentified (but perhaps Burchard), Dodge, Word (cooking, in front), Kolb, Birdseye, Moore (?), and La Rue. The boats (left to right) are the *Glen, Grand, Marble,* and *Boulder.*

Lewis Freeman photograph, Grand Canyon 330, courtesy of the U.S. Geological Survey Photographic Library.

once or wait for the flood to pass. As the boats were already untied, it was decided to go ahead. It was our most spectacular run in some respects—thunder, wind and the rising tide of the river adding to the effect of a hundred yards of blazing driftwood casting its lurid glare upon the head of the rapid. For the first time with bad water ahead, we ran fairly close together. All pushed off at once and then lined out at intervals of about 150 feet. Kolb hugged the right side of the V and made a good run. Lint pulled inside of the big rock at the head of the V and did still better. I ran about Kolb's course, slopping in a few inches of water owing to the *Grand's* way of burying her head like a frightened ostrich. All four boats were in the rapid at once. Running on round the lower curve of the crescent, we found the engineers waiting with an unusually hearty welcome—due to the fact that food and drinking water were in our boats. The night is clear and unthreatening for the first time in several days.

LA RUE: At noon the river rose about one foot in ½ hour. Great masses of drift appeared in the river. Probably a sudden flood from the Little Colorado. The water in the river is so polluted with mud that a bath to any advantage is impossible. Late in the afternoon we came to another rough rapid [Nevills Rapid at 75 Mile Canyon] so camp was made at the head of it. Col. Birdseye's toe is worse, badly infected. Guess he can't climb out when we reach Hance Trail tomorrow. Emery Kolb is having a treatment for lumbago tonight and also H.E. Blake.

KOLB: Frank the cook is feeling badly. He is talking about leaving at B.A. [Bright Angel Creek, Phantom Ranch] suffering with his eyes. Blake seems to have hay fever. Good pools of drinking water in canyon. We take half baths. La Rue stages Freeman, Blake & others in bathing up a side canyon.

LINT, REPRISE: There seems to be bad blood up between La Rue and Freeman for the former told the cook that he had recommended Freeman and that he now wished that he hadn't done so. He said that Freeman wasn't anything except a flunkey and was the laziest man that he had ever seen. He also said that he was going to jump Freeman about it when we got to the Little Colorado River but as yet we haven't seen a scrap and we are several miles past the Little Colorado River. Freeman told Emery that La Rue was yellow, called him several names not fit to write here and also said that La Rue termed himself a technical man and wasn't supposed to do any work. So there you are—gentlemen of the jury.

BLAKE, LETTER:[5] I only remember vaguely Emery's call down of Freeman when he [Freeman] claimed he had the biggest load. It seems to me that it was just prior to the time Col. Birdseye called me into his tent and asked me what I thought of Freeman. I said rather shortly, "His muscles are too slow." The Colonel then said, "I think his brain is too slow," or words to that effect. I believe that if I had said anything drastic against him he would have been sent out by the Grand View Trail.

August 16, 1923 —32,000 ft³/s

LA RUE: The river dropped about 2 ft. during the night leaving our four boats high and dry on the bank. Had to use rollers to get them back into the river, all of which was movied. I also moved the

---

5. Elwyn Blake to Otis Marston, December 18, 1947; box 21, folder 6, Marston Collection. If Blake's memory twenty-four years after the fact is correct, this is the only negative comment about another crew member attributed to Colonel Claude C. Birdseye other than those he wrote about Kolb and Word.

drinking and bathing pools in the side canyon. Moore and Blake were drinking when Freeman jumped into the pool for a bath.

FREEMAN, DIARY: Lint and Kolb had run the rapid [Nevills Rapid] when Blake and I came back to camp. Both of us had a rough passage. I went through down the V, pulled to the left and missed the worst of the first combers. Then I was drawn into the middle of the rapid and at one point hit the heaviest single wave I have yet met. I managed to meet it stern first, but it came over solid and landed full on my back and head, which I had doubled down between my knees to prevent being slammed against the back of the cockpit. The latter was about a quarter filled and the boat swung round and met the next wave side on, shipping another rush of water. Then I swung her stern on again and pulled out to the side of the biggest waves. After bailing, I took La Rue and Moore aboard, and gave the former a heavy sousing when I was down into the middle of a small rapid below.

BIRDSEYE: Found shale under a proposed dam site just below this point [Nevills Rapid] so made no survey. Reached the foot of Hance Trail at the mouth of Red Canyon at 10 A.M., and found it to be 15.3 miles below the mouth of the Little Colorado River. Checked on the B.M. with an error of +8.2 feet, which we considered justly good for a vertical angle stadia line 76 miles long with many long sights on which we had to use the micrometer attachment to measure distances. I had been hobbling along with an infected foot for four days and was surely glad to reach a plot where it could be given attention. Emery Kolb acted as surgeon. Seldom has anything pained as much but I had immediate relief.

KOLB: [As they approach Hance Rapid] I see what appears to be people walking in the sand, but upon arriving it proved to be three cranes riding logs around a huge whirlpool. Set up quite a permanent camp as Col. Birdseye's foot's infected from bad corn on little toe, is in bad shape. After lunch I open medacin [sic] kit and start to work. After steralizing [sic] lance needle etc. I inject novacain in three places then open the wound taking out the puss [sic]. After dressing the foot, Col. took a little smack of liquor but was very nervy.[6]

FREEMAN, DIARY REPRISE: The party has a pretty high percentage of ailing ones at the conclusion of our first stage. Besides Birdseye's toe, the cook is having much trouble with his eyes and back. Kolb and Lint are also complaining of backs. The former's is rather

---

6. Someone filmed Kolb lancing Birdseye's toe. Birdseye smoked a cigarette throughout the procedure, grimacing but slightly. Emery Kolb, untitled videotape, reel 4, Cline Library.

serious. Blake has a persistent cold and headache, and Burchard has a bandaged arm—sunburn. The three members of the party on the *Grand* are the only ones so far not reporting ailments. Probably this will not last long.

LA RUE, REPRISE: Put up the radio and listened to the bed-time stories and a fine program from 8 to 10. Moore and I listened to the radio until 10 o'clock and I was just going to bed when he came to my bunk and asked me to come up to his bed in the rocks for he thought he heard a rattle snake. I put on my shoes and with a flash light I climbed over the rocks to his bunk. A rattler sure enough. With the flash light I found him between the rocks about 1½ feet from the head of Moore's bed. I tried to pin him to the rock with a stick but he got away. It was sure a big one. It is hardly necessary to say that Moore moved his bed down on the sand where he should have put it in the first place. Our noise and the light woke up the Col. His bed was also in the rocks not 30 feet away. I told him I had just tried to kill a big rattler but he got away. The Col. didn't sleep any more that night. Every noise he heard sounded like a rattle snake.

BLAKE: [At Hance Rapid] ... immediately starting making camp, as it was thought best to get new oar locks made before running the rapid. Emery, Leigh and I walked to the top of the rim, starting from camp at 3:30, and arriving at the old Canyon Copper Co.,[7] diggings about dark. Dodge had started with us, intending to go far enough to get a good view of the canyon, but turned back after going about a mile and a half. We found a spring of good water below the mine, and filled our canteen. The trail from the mine to the Horseshoe bench is mostly shot out of solid limestone and is exceedingly steep, and is out of repair. On arriving on top of the bench we discussed the advisability of laying out all night but decided to go on until a good sleeping place was found. Two thousand five hundred feet of steep trail lay ahead of us and each felt that they had already done a good day's work, but all plodded upward in the dark, stumbling over loose stones and slick rocks. As the hours wore on we could look back at each resting place and see that some progress had been made toward Grand View Point which rises five thousand feet above the river and is 7,500 feet above sea

---

7. In 1901, the Canyon Copper Company purchased, from Peter Berry and Ralph and Niles Cameron, the Grandview and Last Chance copper mines, which Berry had located eleven years before. William Randolph Hearst acquired the property in 1913; mining ended in 1916. Billingsley, Spamer, and Menkes, *Quest for the Pillar of Gold*, 65–67.

level. Several times one or the other of us fell instantly asleep when a rest was called, but at last the top was gained.

KOLB, REPRISE: Had it been light enough to see the shacks we would likely have stayed at the mine but we slowly dragged up the hill. Leigh would have laid down and slept any where but Blake & I encouraged him on. Before the top was reached I was ready to quit myself, went to sleep on the trail twice while resting. Had a little rye crisp to eat. We reached Dick Gilliland's[8] at 12 midnight. They let us in after the dogs are put away. Don't blame the dogs for wanting to devour us.

August 17, 1923 — 28,800 ft³/s

KOLB: About 7 am. Dick comes in and asks us to breakfast. What a breakfast. Sliced peaches with real cream, bacon, fresh eggs, three cups of coffee. Dick takes us to El Tovar in his Buick. We have another breakfast at home. In the evening we invite Leigh & Blake to supper, also Shirley girl.[9]

LINT: Blake and I took in Ed Kolb's show and appreciated it very much. Mrs. Kolb got up a big dinner to celebrate my birthday and we enjoyed it very much. Afterward Edith played the piano and the rest of us tried to sing. Roger Birdseye got into Grand Canyon with the supplies at 9:30 P.M. Edith took Blake and me out in the car and showed us the sights from the rim.

BLAKE: Although it was not my birthday I received a watch fob in remembrance of the Lee's Ferry days.

FREEMAN, DIARY: Birdseye's toe is improving. He has discarded his crutch and made a shoe out of a gas mask bag.

LA RUE: Ran survey down the river without the boats. Could go but a short distance following the water. The granite Gorge starts about ½ mile below Hance Rapid. By following the rising formation we carried the survey to Mineral Canyon where an excellent dam site was found. About ½ mile below Hance Rapid on left bank we found a boat cached under the rocks well above high water. The boat was a sectional steel boat with flat nose fore and aft. Boat and oars nearly new.[10] At Mineral Canyon we made plans for a detailed survey and returned to camp about 2 P.M.

---

8. Dick Gilliland was apparently the caretaker of the closed Grandview Hotel. Blake, "As I Remember," 165.
9. Eleanor (or Elinor) Shirley joined Blake and Lint on their Grand Canyon picnics.
10. We have been unable to find any specific information about this boat, but miners probably used it to cross the river to access Hance's asbestos mine.

August 18, 1923 — 30,200 ft³/s

LINT: Left for Grand View at 9:30 with Roger and Emery followed later with his family and Blake. The mules had been sent out earlier and as soon as we got there we packed up. Mrs. Gilliland and Edith Kolb decided to come down to the river and watch us run the Hance Rapid. Other members of the party were Charley Fisk, Roger Birdseye, and the packer. Blake and I left on foot at 11 o'clock and reached the river at 3:15. 15 miles in 4 hours and 15 minutes. The pack train arrived at 5:25.

KOLB: Our row locks are made by both blacksmith and garage. The oars are sent to Flagstaff instead of Grand Canyon. We wire for them and leave their bringing for Ellsworth who is coming down to see us run the rapid.

FREEMAN: I walked up the trail to the crossing of Mineral Creek, and then followed the latter to the foot of the cliffs of the outer walls. At one point on the trail I had a good view down the river to the Sockdolager Rapid. Although over a mile distant it showed unmistakably as heavy water, especially at the head, where it drops almost out of sight, so great is the fall. However, our immediate business is the passage of the Hance, which probably has more power for evil than the much-written-about Sockdolager.

LA RUE: Made detailed survey of Mineral Canyon dam site #4. Survey made from the walls although we were able to climb down into Mineral Canyon to where it drops off sheer for 30 ft and 200 ft further down there is another sheer drop of about 50 ft. About 3 P.M. L. Lint and Blake passed on the trail above us returning to camp from Grand View and El Tovar. They announced that the pack train would be down in about two hours and that ladies would be in camp. They also warned us to put on some clothes and cut out all rough language. About 5 PM the pack train arrived with supplies and mail. We had some feed. Real potatoes and fresh fruit. Everybody soon settled down to read letters and newspapers. Just before the pack train arrived a near cloudburst hit us. We collected four buckets full of water and half doz. pans full from the canvas fly over the cook outfit.

August 19, 1923 — 33,000 ft³/s

BIRDSEYE: Use the pack train to portage all of the equipment to a sand bar on the left bank below Hance Rapids. This Rapid is one of the worst on the river. Has a fall of 28 feet in a few hundred yards and is full of rocks over which the water pours in monstrous waves. Kolb ran first with the *Marble* as is always the case and carried

Lint as a passenger. Lint loves rough water and wanted a close view of the rocks before running through himself. Lint followed in the *Boulder* with Edith Kolb as a passenger and came through in fine shape.[11] I was below and did not know she was going through and confess I was surprised. However, there is little danger if the rocks are avoided. She is surely a game girl and guess she had a good thrill. The other boats all made the run in fine shape.

Kolb: Edith coaxes me to ride the rapid so I leave her go with Leigh.

Lint: Edith … rode on the stern hatch cover of my boat. Blake followed me about 200 feet so that he could help in case I had any trouble. I took on about 5 inches of water in my cockpit but otherwise we both got through O.K. Freeman got a poor start but he had his usual luck and got through O.K.

Blake, autobiography:[12] Lint pulled out into the swirling current. In moments the boat was rearing and plunging, with Lint straining to avoid protruding rocks. He managed to pull in a little below the camp we had established near the foot of the rapid. "I never came so close to capsizing in my life," he told me. Edith must have had a moment of fear, as the *Boulder* heeled over, almost on its side, but she never showed any nervousness after her wild ride.

Freeman: All loads were removed of the boats and they were taken out on the beach and carefully gone over for injuries from previous navigation [prior to running Hance Rapid]. A line was run along the bottom of each, with just enough slack for a man to hold to in case of an upset. I wound the handles of my oars heavily with rubber tape, as I have had a good deal of trouble in gripping them after my hands were wet. The number of large boulders cropping up all through the head of the rapid, and a bad hole half way down in the middle of the current, limited the possible courses to two—one on either side of a large rock showing just above water at the very brink of the first fall. The important thing was to keep to the left of the big wave below, as going into the hole above it—the

---

11. By riding through Hance Rapid, Edith Kolb became the first woman to ride through a major Grand Canyon rapid. It also should be noted that her boatman was Lint, who was developing a romantic involvement with the young woman. In film footage of the expedition, Emery Kolb is shown securing Edith's life jacket and seeing her off. Kolb filmed his daughter's historic ride through the rapid, as noted by Freeman in his original diary. This footage is on Emery Kolb's videotape, reel 4, Cline Library. The incident was significant enough to warrant mention in the Flagstaff newspaper, which noted, "this young woman seems to have inherited the intrepid part of her father and uncle." *Coconino Sun*, September 21, 1923.
12. Blake, "As I Remember," 166.

When Edith Kolb joined Leigh Lint in the *Boulder* to run Hance Rapid on August 19, she became the first woman known to run a major Grand Canyon rapid. Here she poses with Lint, at left, and her father Emery Kolb, apparently near Hance Rapid.

Emery Kolb Collection, NAU.PH.568-803, courtesy of the Cline Library, Northern Arizona University.

---

bottom of which we could not see from the highest vantage—meant a certain upset if not a smashed boat. Kolb inclined to favor a course just to the left of the mid-stream rock, then pulling hard to the right to avoid two or three nearly exposed rocks just below, then a heavy pull to the left to miss the big wave and make a landing on the beach below the new camp at the foot of the rapid. To me it seemed simpler to run to the right of the dividing rock at the head, and then make one long, hard pull of it away from the middle wave.[13] A log passing close on that side of the rock I saw go into

13. Freeman's assessment of Hance Rapid follows the route favored by many Grand Canyon boatman at this water level—an entry right of the rocks in the center of the channel, a move to the left through a small pool, which some call the Duck Pond, and a line up to avoid a series of waves and holes to the left and right in the rapid.

the middle wave. This meant that sharp pulling, with no bad luck, ought to clear that most troublesome obstacle. At the last moment Kolb decided to run this course. Kolb ran first, borrowing my 9-foot oars to increase his pulling power. The *Marble* was put in about 25 feet to the right of the boulder at the head, from where Kolb began pulling to the left. It was a hard fight in the rough water, and his boat got a rough slap from the side of the big middle wave in passing. Lint says he hung over the hole and could see way down into the bottom of it. Dodge was waiting in the water at the foot of the rapid to throw a line to help in landing, but Kolb could not get near enough to allow Lint to reach it. He was carried down the next riffle and did not pull in to the bank for over 200 yards. From here we lined his boat back up stream to the beach below camp. [After Lint's run] Ed [Ellsworth] Kolb arrived with the new 9-foot oars, and Blake took a pair of these for his run. He went through well and landed beside the *Boulder*. The difference in the way the empty boat pulled and handled was evident as soon as I pushed off. With more confidence in my ability to control the boat, I ran closer to the midstream boulder than I otherwise would have, passing an oar's length to the right of it. This left me well set for my pull away from the big wave below. Wave after wave that would have swept the loaded boat from end to end, she now lifted over buoyantly. Shipping no more than spray, I passed the big wave and hole by about 15 feet and pulled into the beach after a comparatively dry and easy run. Ours is probably the first party that has run all its boats.[14] We went over our loads carefully this afternoon, eliminating all that was superfluous in the way of outfit and provisions. This will be sent out by the packtrain. As all passengers will have to be carried through at least two rapids of the Granite Gorge, we want to be no heavier than is necessary.

August 20, 1923 — 33,200 ft^3/s

BIRDSEYE: Called this day Sunday and took another day's rest, as the boys who had been to the river were still pretty tired. On all five nights at Hance Trail (16–20) the radio worked fine and heard KHJ every night.

KOLB: Eddy Bruce the guide lets three of his mules get away from him. He and Roger with Fisk continue after breakfast. At noon Bruce comes back with the mules. I help him pack. Worked

---

14. The 1923 USGS expedition was unaware of George Flavell and his successful run of Hance Rapid in 1896.

The day after running Hance Rapid on August 20, Lint discovered gaps in the bulkhead of the *Boulder* that required repair. The three older boats—the *Boulder*, *Glen*, and *Marble*—were all extremely fragile and fractured easily. The man at the far left, seated on the Grand, is probably Moore; La Rue is sitting by the stern. The men working on the *Boulder* are Dodge, Blake (?), Freeman, and Lint.

Emery Kolb photograph, Grand Canyon 51, courtesy of the U.S. Geological Survey Photographic Library.

---

all morning on changing sockets for 9 ft oars. About 2 P.M. Blake Leigh & I go up Red canyon about four hundred ft. Snooze & write. As I look up I see the leaping waves of mud at the beginning of the rapid but the boats are at its end.

LA RUE: Played bridge in PM Col. Birdseye and Moore against & Burchard & me. Wind and little rain broke up the game.

August 21, 1923 — 33,400 ft³/s

BLAKE: After we had put the boats in the water Leigh found that the bottom of the *Boulder* had drawn away from the forward bulkhead so that water could enter the cockpit, so the *Boulder* was hauled out of the water again and a number of screws were put through the bottom to draw it up. All three of the Edison Co. boats are pretty rotten and only good luck can make them last the trip through.

BIRDSEYE: After carefully reloading the boats we entered the granite gorge where we anticipated trouble in getting footing for instrument and rod stations on account of the sheer granite walls and swift water. Passed #4 Dam Site at mile 16.5 and completed the detailed survey by getting the water level and width of the river at the site. Ran through very swift water with walls only about 125 feet apart and had considerable difficulty in planning our sights. Reached the head of the dreaded Sockdolager Rapids at mile 17.2. Dellenbaugh describes this as an 80 foot fall in ⅓ mile with waves 30 feet high. We found the 30 foot waves but the fall was only 19 feet, most of it in the first 100 yards. It is impossible to climb around this rapids so all had to ride. Kolb and the cook went first in the *Marble*. They were out of sight half the time but went through in nice shape. Blake and Dodge on the *Glen* followed and made a good landing below with only a 2100 foot foresight. (Had expected to have to use the micrometer for distance). Freeman with La Rue and Moore went next in the *Grand* and took practically no water. Lint with Burchard and I brought up the rear and all got a good ducking. Nearly broke my neck looking up from the rear deck to see the tops of the on-rushing waves. We all got a thrill but think the danger of the rapids has been magnified by previous explorers.[15] The waves are enormous but the channel is free from rocks so the only danger is from upsetting. We continued the survey with some little difficulty to find proper instrument and rod stations and camped on the left bank opposite the mouth of Vishnu Creek at mile 19.7 [Grapevine Camp]—nearly on the west edge of the Vishnu sheet.[16] Set up the radio with aerial in a N.W. and S.E. direction. Heard San Francisco (Hales) for the first time—also KZN at Salt Lake and of course KHJ at Los Angeles. There was little static and all came in fine. The *Literary Digest* article of about July 15, claiming it impossible to get radio communications in a deep gorge has surely been disproved.[17] The gorge here is over a mile deep and the walls only 150 feet apart at the river.

FREEMAN: Moore says that he has observed little true granite in the gorge so far—meaning, I take it, the speckled kind they ship from Vermont for grave-stones. There is gneiss, schist, feldspar and other ancient igneous rocks, however; but no matter how distributed the effect is rarely dark and gloomy as we had been led to expect.

---

15. Birdseye reflects on something one of his predecessors, Robert Brewster Stanton, had already determined: Powell had greatly exaggerated both the drops and the dangers of river expeditions in Grand Canyon; Webb, *Century of Change*; Webb, Belnap and Weisheit, *Cataract Canyon*.
16. Freeman noted that this was "the first sand-bar we have seen in Granite Gorge."
17. "How Far Can My Radio Set Receive?" 25.

LINT: Something must have happened for Freeman has done a little work in the last three days.

August 22, 1923 — 27,900 ft³/s

BIRDSEYE: Immediately after starting passed the mouth of Grapevine Creek coming in from the south and ran into Grapevine Rapids (#7) at mile 20.0, with a fall of 17 feet in a few hundred yards. This is considered a bad rapids with many rocks but we ran it easily and did not get as wet as in the Sockdolager. One can not climb around this rapids, in fact there are few places in the Granite gorge where one can travel much along the river bank. One can climb up to the Tonto rim in several places but it is 1000 feet or more above the river. We have been running on a fairly high stage of water which undoubtedly makes the rapids safer to run especially those which contain rocks near the mean water surface but the high stage makes the waves break higher and makes larger eddies and whirlpools below them. Ran #8 Rapids at mile 22.1, a small one with a fall of only 6 feet, but with many rocks on the south side. At noon we passed the mouth of Clear Creek coming in from the north at mile 22.7. This stream flows fine clear water and enters the river in a narrow blind gorge the presence of which one does not suspect until he is actually at the mouth. In fact none of the streams in the Granite Gorge cut down in wide valleys near the river and one can easily pass any of them unnoticed. It is very dangerous to land right at the mouth of Clear Creek but one can effect a good landing in a cove about 200 yards above the creek and by a rather difficult climb over the rocks can get into the creek for water. The Kolb brothers have ascended the river in a canvas boat from Bright Angel to Clear Creek and traveled up Clear Creek several miles from the river. Surveyed Dam Site #5 about 1200 feet below Clear Creek at mile 22.9 and camped at this point on the south bank. The south walls of the dam site are good but the north wall slopes back at an angle of about 45°. The river is 200 feet wide and a 300 foot dam would be about 900 feet across at the top. The site has good prospects of a spillway from a point about ¾ mile up Clear Creek by tunnel into the next small canyon to the west.

KOLB: River fell in night about 2 ft. Awakened early by gnats. Engineers land on S. side of head of rapid [Grapevine Rapid]. Bad rocks on north. I ship but ½" water. From 15 ft above water line, at end of rapid, the boats would be 3 to 5 second out of view between waves. Swift water soon brought us to Cottonwood Canyon where

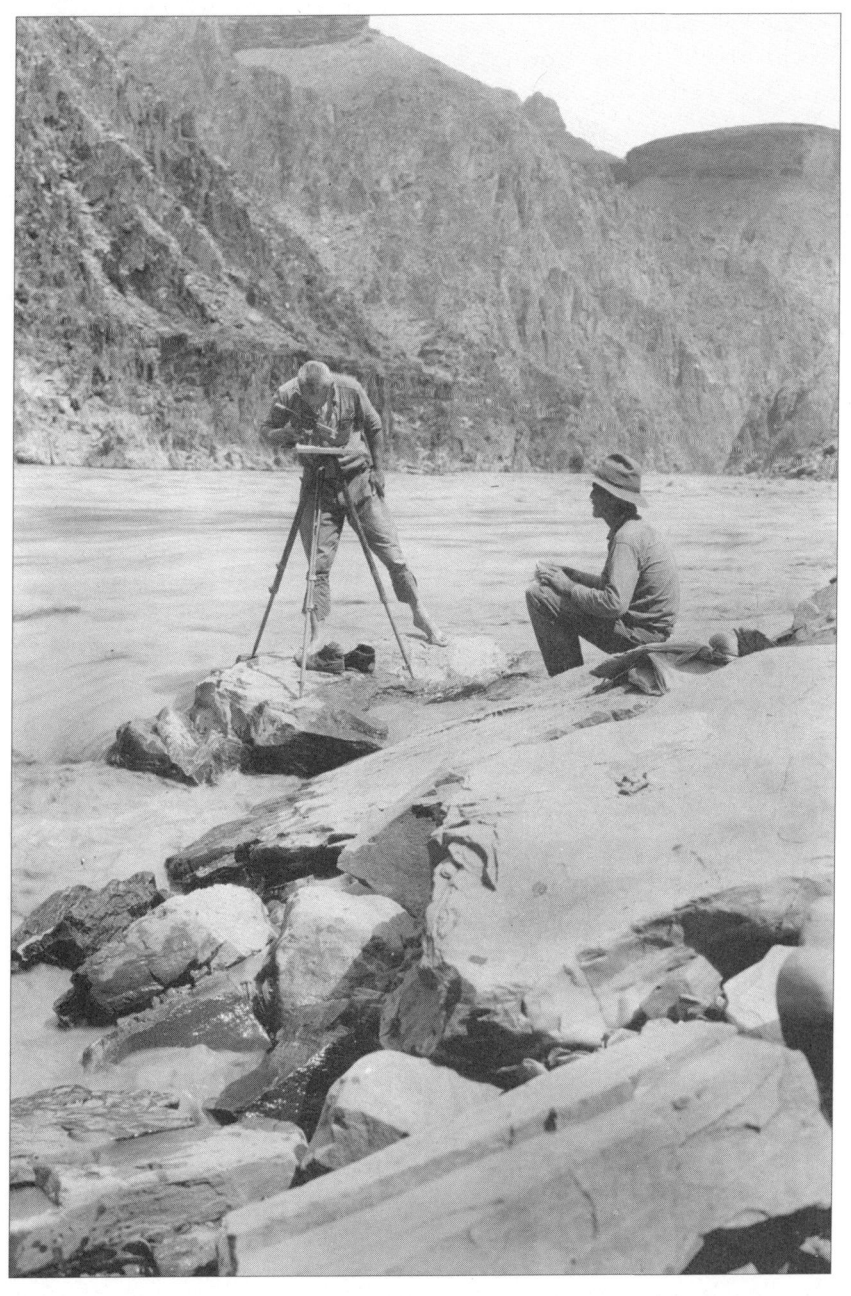

Claude Birdseye records while Roland Burchard surveys in the upper Granite Gorge in August.

Lewis Freeman photograph, Grand Canyon 318, courtesy of the U.S. Geological Survey Photograph Library.

Ashurst was killed.[18] I try to land at Clear Creek but can't. Land on bar just below. Frank and I get water from canyon by climbing 100 ft, ¼ mile. Freeman almost did not make it. La Rue has a cross section made. Frank tells the Col. that Emery remembers no camps between Clear and B.A. so we camp here.[19]

BLAKE: About one o'clock lunch was eaten on the only sandbar of any size seen today.[20] A cross section for a damsite was made after lunch, and La Rue took a movie of the surveyors at work (in life jackets right near the terrible river). It was decided to camp on the sandbar rather than risk having to sit up all night if we went on, camping places are so scarce. Several of us went swimming, Dodge leading the way by plunging in with all his clothes on (consisting of a pair of overalls cut off above the knees.) Birdseye, La Rue, Moore and Burchard played bridge this evening …

LA RUE: Radio best ever last night. Aerial high but in NW direction. Got KHJ at Los Angeles, the Deseret News at Salt Lake and Hale Bros, San Francisco. "Onward Christian Soldiers" from Hale Bros. "Where is my Wandering Boy Tonight" from Deseret News. The phonograph music from S.L. came in like a brass band. KHJ was strong as usual.[21]

August 23, 1923 — 24,800 ft³/s

BIRDSEYE: Ran #9 Rapids [Zoroaster Rapid] at mile 23.2 which had a fall of 7 feet. Surveyed Dam Site No. 6 at mile 23.8. Ran #10 Rapids [85 Mile Rapid] at mile 24.0 with a fall of 5 feet. Surveyed Dam Site No. 7 at mile 24.7 [Cremation Creek], and considered this the best dam site section we had seen, so far as narrow section and good abutments are concerned. There is, however, no natural spillway and no suitable power site except by blasting out in the site of the canyon walls. Reached the Bright Angel gaging station at mile 25.8 and at 1:00 PM we came to the mouth of Bright Angel Creek at mile 25.9. The water elevation at the mouth of Bright Angel Creek

---

18. William H. Ashurst, 57 years old, was prospecting in the Grand Canyon near river mile 87 when he was killed by a falling slab of Vishnu Schist in January 1901. His son, Henry Fountain Ashurst, was one of Arizona's senators from 1912–41, who, with Carl Hayden, introduced legislation that would establish Grand Canyon as a national park. Michael P. Ghiglieri and Thomas M. Myers, *Over the Edge: Death in Grand Canyon* (Flagstaff, AZ: Puma Press, 2001), 274.
19. This is Kolb's last entry. In his diary, he wrote the date "Aug 23" but nothing beside it.
20. The scarcity of sand bars in Upper Granite Gorge was noted by other river runners who experienced the unregulated Colorado River. Some who run the regulated river do not realize that the absence of sand bars is not just caused by the operations of Glen Canyon Dam.
21. Like Kolb, La Rue stopped writing in his journal on this date in spite of having plenty of blank pages. Both men may have sent the journals home.

From left to right, Leigh Lint, Emery Kolb, and Elwyn Blake pose for a photograph, probably taken at the Kolb Studio by Ellsworth Kolb.

Courtesy of George Lint.

was found to be 2435 [ft]. Kolb, Lint, Blake and Dodge ran on in the *Marble* to a point near the mouth of Pipe Creek and then walked up the trail to Grand Canyon on P.O. The rest had supper at Phantom Ranch,[22] a Harvey resort about ½ mile up Bright Angel Creek. Freeman, Moore and Word started up the trail at 8 P.M.... Burchard, La Rue and I camped at the hydrographer's home near the mouth of Bright Angel Creek and prepared to walk up the trail the next AM.

FREEMAN: We found Donald Dudley in charge of the USGS station, and stored our beds and dunnage there to remain during our

22. Phantom Ranch, designed by Mary Jane Colter, was completed in 1922. Elizabeth J. Simpson, *Recollections of Phantom Ranch* (Grand Canyon, AZ: Grand Canyon Natural History Association, 1984).

absence at El Tovar. Dudley informed us that the river was flowing over 25,000 second-feet, and had been up as high as 34,000 [ft^3/s] during the previous week. It was at about its highest during the days we were running Hance and Sockdolager rapids.

BLAKE: Emery, Leigh, Dodge and I took Emery's boat and crossed the river a half mile below Bright Angel Creek. We then pulled the boat out on the sandbar and tied it securely, after which we started over a secret trail which Emery had marked out and which cut off about four miles from the distance traveled by trail from the bridge. There is no wonder that the trail has never been used by any but the Kolbs, as very few persons would have the courage to use if they knew of its existence, as it is over granite ledges and along precipices several hundred feet above the river where there is no footing except small projections a few inches wide which make the use of both hands and feet necessary in traveling the short cut. In about forty minutes of climbing we came to the foot of Bright Angel Trail and were soon on our way toward the rim. About a half mile from the river a cold stream tumbles down the granite wall on the right, forming pools and waterfalls. We were guided by Emery to the "Bathtub" a short distance from the trail, where we took a plunge in the cool, clear pool in the solid granite. This bathtub, tho it is very near the much traveled trail, is very little known, as are the two others hidden by willows and cattails nearer the trail, even by the guides who pass within a few yards of them each day. After the plunge we continued our climb past the "corkscrew" and to the "Indian Gardens,"[23] where we stopped for a drink from a cool spring. We talked a few minutes with the old prospector who lives there[24] and were on our way again. When within a thousand [vertical] feet ... of the top, we heard a call from above, and Emery recognized the voice of his brother. We reached the top at 7:40 and were met by Ellsworth and Edith Kolb who informed us that supper was waiting, so we went to Emery's place and had a fine meal,

---

23. Indian Gardens, with its permanent water supply, is situated 4.5 miles by the Bright Angel Trail from the South Rim. It was originally a Havasupai Indian farming site. Around 1901, Ralph Cameron developed it as a tourist camp, which operated until 1910. Today, the National Park Service maintains a campground and ranger station at Indian Gardens, which is a popular day hiking destination. Brian, *River to Rim*, 71; Anderson, *Living at the Edge*, 86–87, 105.

24. The man they talked with likely was one of the caretakers for Ralph Cameron, senator from Arizona, who had mining claims at Indian Gardens; Billingsley, Spamer, and Menkes, *Quest for the Pillar of Gold*, 64–65. Later that year, park rangers removed the caretakers and assumed possession of the property.

after which Dodge, Leigh and I got a cottage[25] containing four beds where we stayed for the night. Dodge using the covers from the extra bed on his own as the weather was somewhat cooler than at the river.[26]

August 24, 1923 — 22,300 ft³/s

BLAKE, AUTOBIOGRAPHY:[27] Emery came along as we ordered breakfast at the Bright Angel Restaurant. As we sat at the counter, another man came in and sat down beside Emery. They exchanged greetings, after which Emery turned to us and said: "Fellows, meet the governor of Arizona." Dodge thought it was a gag of some kind. He got off his stool, bowed in mock deference to a dignitary, and made a frivolous acceptance of the introduction. He was somewhat taken aback when convinced that it was in truth Governor Hunt,[28] of Arizona. The fact that the governor was eating breakfast at the Bright Angel Restaurant, instead of at the El Tovar,[29] had made Dodge question his identity.

FREEMAN, DIARY: Pretty nearly killed Word making the rim, but finally landed him intact.[30] Breakfast was a joy. Also a good morning sleep, followed by lunch and later by dinner. Birdseye and Burchard arrived by mule in time for lunch and our bags arrived from Flagstaff in time for a change before dinner. El Tovar has made us very comfortable, especially Moore and me, who appear to have a section of the Bridal Suite in the second floor corner. Roger Birdseye and wife drove over in time for dinner. Amusing story in the papers today about a carrier pigeon being picked up in N. York with plea for aid from Grand Canyon Survey party!

---

25. Blake, Lint and Dodge probably stayed at one of the Bright Angel Hotel (now the Bright Angel Lodge) cabins nestled near the Kolb Studio on the South Rim.
26. Blake may have found Dodge's bedding needs noteworthy because Dodge used only minimal bedding on the river. Dodge had written, "I gave my Abercrombie and Fitch sleeping bag that Birdseye furnished, to the cook. Those early bags were built for vertical rain only and had too much red flannel sewed in them for August weather anyhow. I was tough then, so all my bedding consisted of was an 8 x 8 foot canvas; going almost naked, it meant that when I jumped ashore, I was practically landed for the night." Dodge, *Saga*, 37.
27. Blake, "As I Remember," 169.
28. George W. P. Hunt was the first governor of the state of Arizona and served from 1912–17, 1917–19, 1923–29 and 1931–33.
29. El Tovar was, and still is, a fancier and pricier restaurant than the Bright Angel. The same is true for the corresponding hotels. It appears that the boatmen and Dodge were housed at the Bright Angel, while Freeman and the scientists stayed at El Tovar.
30. Curiously, Word hiked back down to the river the next day.

BLAKE: Mrs. Kolb drove us, Leigh and I, out to the rim at various interesting points. La Rue was the only one who stayed at the river. At 4:15 the party were the guests of the Kolb Bros., who showed their trip from Green River, Wyo., to the Gulf, but showed two and a half reels of the Cataract Canyon and Glen Canyon trip of 1921.

BIRDSEYE: We plan to remain here until Sunday August 26, when Mr. Stabler is expected to join our party. We expect to continue our river trip on August 27. Expect to arrive at Hermit Trail about August 30, at Bass Trail about September 3, at Havasu Creek about September 10 and Diamond Creek about October 7. Hope to be able to join the old Burchard survey above the Grand Wash about October 15th.

August 25, 1923 — 20,600 ft³/s

LINT: Tourists are sure a bone-head bunch as they wake a person up about 4.30 A.M.

BLAKE: I got my suit case from the warehouse and had my suit pressed. Leigh and I went with the Kolbs to a dance. I danced four or five times and came home before midnight.

FREEMAN, DIARY: More hearty guzzling of food. Wrote letters most of day, and read proofs of my Colorado book.

BURCHARD, LETTER: Dear Kate and Harry,[31] We are this far on our way—we arrived day before yesterday at Suspension Bridge [where the Black Bridge is today] at foot of Bright Angel trail and yesterday we all came to the top for a change. This is a great place and a great change. Everything is run by the Santa Fe and Fred Harvey which means it is well done—and expensive—Tourists both auto and train are numerous in fact it will be a record breaking year for the park.—The elegance of this fine hotel is quite dazzling to one who has inhabited the river for thirty days. The climate is also a pleasant change—quite chilly here and quite hot in the bottom altho not as hot as Boulder Canyon or Lee's Ferry We have finished about 90 miles of our 250 mile trip. So far we have had no troubles. Our type of covered over boats with only one open cockpit for the boatman seems the correct style as the water tight compartments would keep the boats afloat even with a hole it and the cockpit full—We lost a little open boat which the Rodman used for a run-about in Rapid #13 since that he rides in one of the boats. The Fox news had a

---

31. Kate and Harry were Burchard's sister and brother. The letter is written on El Tovar stationery. Roland Burchard to Kate and Harry Burchard, August 25, 1923, Burchard Collection, box 1, folder 1, Cline Library.

camera man to meet us at Suspension Bridge for some movies but we had already camped when he arrived so he is waiting on the river for us to start again which we expect to do on Monday the 27th. If we have no trouble and not too much work we will likely tie in to our old line about Oct 15 or 20. Our trip thru Marble Canyon was quite pleasant but uneventful. Our radio works very satisfactorily altho we have failed to get any personal messages except official. It was quite remarkable that altho in the depth of Marble Canyon facing our most thunderous Rapid—The Soap Cr Rapid—The Waterloo of several expeditions we learned of Harding's death 45 minutes after it happened. It came as a jolt as we had not known of his former ill health. We have run about 27 rapids of about the same magnitude of our last one and several perhaps 5 much worse as regards flow and fall—namely Badger Cr—fall 13 Soap Creek around which we portaged and even carried the boats around the worst stretch of 150 feet—which took all day fall 16 feet—Hance Rapid which required portageing [sic] the loads and running the boats empty fall 28 feet in about mile continuous rapid "Sockdolager" a famous rapid—not far above here—Major Powell reported 80 foot fall turned out to be 18 narrow channel and smooth sides, very high waves—All other rapids are 10 feet or less and altho they are nearly continuous in the Granite Gorge at this stage of water, which is quite high. They have given no trouble to amount to any thing of course the skill of our four boatmen Emery Kolb Captain accounts for our good luck our luck altho good does not compare to Chaddock [Chaldock?]— we have a wise geologist who has quite a reputation—Mr. Birdseye records for me mostly altho of course he could do the topography better and faster than I. La Rue the Hydrog. Eng. has a movie camera also Emery Kolb. One of the boatmen [Freeman] is a writer of some note. A Mr. Stabler of the Washington office (Hydrography) will join us here making eleven. Letters from Eleanor seem to indicate that she is enjoying her new family. I was sorry not to have a letter from you here. I would very much to know how you are getting along. We will get mail from here for about 2 weeks and I hope you drop me a line. With much love, Roland.

August 26, 1923 — 18,800 ft³/s

BLAKE: The Kolbs drove Leigh, Miss Shirley and I out into the timber near the river east of town where we had a picnic lunch. We returned to town about 5 o'clock. Edith drove Miss Shirley home and as she returned was arrested for speeding.

FREEMAN, DIARY: Still eating. Mrs. La Rue came in on a surprise visit to her husband this morning, and left with the pack train for Phantom. Had a chat with Weeks of the Arizona Gazette, who is here with the Governor's party, and gave him a story for the A.P. [Associated Press newswire]. Moore, Burchard and I went down to Phantom on foot this afternoon.

MABEL LA RUE TYPESCRIPT:[32] ... when I received a letter from Mr. La Rue [her husband] saying that he would be at Bright Angel Crossing for several days, I decided that never again would I have such a good opportunity to see the Grand Canyon. It seemed fitting that the wife of an engineer, who is supposed to be a Colorado River expert, should have seen something more of the river than can be seen from a train window at Yuma. So I packed a few things and boarded the train for El Tovar. On my arrival there, I recognized one member of the survey party [Freeman], and he told me that my husband was the only one of the party to stay down on the river. I could talk to him by telephone, so I called him up. He thought I was safely home in Pasadena and refused to believe that I was really talking to him from the hotel, even when I told him so. I had to give my street address in Pasadena and even name the adjacent streets before I could convince him that he was not being made the victim of a joke. And then, instead of coming up to see me, he asked me to make the trip to the bottom of the canyon. If I had known what was ahead of me I should have refused on the spot, but I knew other people had gone down the trail so I was willing to try. In another half hour I was mounted on Katie's back, ready to begin the descent. I had never been on a mule before, and for the first few minutes I thought my breath would never last for twelve miles. Every step that mule took made me feel as if someone had hit me in the stomach. But after a little I became accustomed to the gait and began to look around me. Then I decided the top of the canyon looked much better than the depths below. I was sure that long before we reached the spot called "Indian Gardens," I should slide over Katie's head and land among the rocks hundreds of feet below. I kept thinking how disappointed Mr. La Rue would be when

---

32. Mabel La Rue, undated typescript, box 10, folder 1, La Rue Collection. Mabel E. La Rue remains somewhat elusive. We found listings in the Social Security Death Index and California Death Index that we believe to be her, indicating she was born in Ohio in 1888, and died in Los Angeles in 1970. Social Security Death Index, RootsWeb.com, http://ssdi.rootsweb.com/cgi-bin/ssdi.cgi?lastname=larue&firstname=mabel&nt=exact (accessed March 17, 2006); California Death Index, RootsWeb.com, http://vitals.rootsweb.com/ca/death/search.cgi?surname=larue&given=mabel (accessed March 17, 2006).

the party reached Phantom Creek without me. Why is it, that when we reach a sharp turn with a straight drop before us, those mules always lean out over the edge to crop a green leaf, when there are some on the inner bank just as tempting? Or else they choose the sharp turns as the proper places to trot in order to catch up with those ahead? But after swallowing my heart a few times, and finding Katie still on the trail, I cheered up. I wanted to ask the guide how far we had traveled, but decided I would wait until I was sure we had at least covered a third of the distance. So, soon after I began feeling a tooth-ache in my left knee, I casually asked, "About how far do you think we have traveled?" The guide looked around, and about as casually replied, "Oh, about a mile and a half." I just about decided right there, to get off and walk back, but I remembered it would be three more months before I could see Mr. La Rue and tell him just exactly what I thought of him, so I took a little firmer grip on the saddle-horn and went on. We reached "Indian Gardens" at noon, and stopped there beside a stream of clear water to eat our lunch. There were some very comfortable benches under the bushes, and in spite of a thunder storm mumbling in the distance, I was happy. There was a chance that Gabriel might blow his horn, or we might be struck by lightning before lunch was over. At one o'clock the guide called "All aboard," and we climbed on our mules and were on our way. The next few miles we rode over a plateau, and I was just beginning to think that the trip wasn't so bad after all, when we came out on top of an awful gash in the earth. The guide pointed out the green roofs of the bungalows at Phantom Camp, our destination, 1500 feet below. "Abandon hope all ye who enter here" flashed through my mind. Dante could have written a more realistic "Inferno," had he spent several days riding around on these trails! It was here that our guide began to tell stories about mules that had slipped over the edge. One slid about a hundred feet down the side of the mountain before it regained its footing. When we were about half way down, he pointed out the four Geological Survey boats the men had used in making their trip into the canyon. They looked as if I could pick them up and put them in my pocket, and take them home to the children. When at last we reached the Suspension Bridge, and the guide said we would have to dismount and wait while he took the mules across, one by one, I began to think that life was worth living after all. Mr. La Rue was waiting for me on the other side of the river, and I was glad of an excuse to walk the rest of the way.

August 27, 1923 — 17,600 ft³/s

STABLER: On arrival found Birdseye waiting at El Tovar. After getting breakfast repacked in 2 special bags and went down trail with pack train. Interested to see that I could readily recognize the formations passed. Reached end of trail at Garden [Pipe] Creek about 12:30. I talked with Birdseye, La Rue, and Moore, getting up to date on what had been done. Just before supper took a little geologizing walk with Moore. Bedded down on sand bank after looking over Pipe Creek dam site. This is good site with indication of bed-rock rapid ⅓ way across—good section—good saddle spillway—excellent abutment and walls. Development would flood suspension bridge, Phantom Ranch, etc.

BIRDSEYE: Burchard carried line from Bright Angel Creek to Pipe Creek ... and checked on the B.M. at mile 27.2 with an error of +1.9 feet (in 12 miles from Hance Trail).

FREEMAN ARTICLE:[33] When we returned to the boats [parked below the Suspension Bridge] ... we found there a cameraman [J. P. Shurtliff][34] from the Fox News Company, who bore an order from Colonel Birdseye for a passage in one of the boats as far as the foot of the Bright Angel Trail. The *Grand*, as rather the most stable of the fleet, was designated for the job. As a motion picture camera set up in the cockpit would have interfered with the use of the oars, there was nothing to do but to open one of the holds and establish tripod and operator inside. As this was the first occasion any of the boats had entered any but comparatively slack water with a hatch removed, I started out with the idea of being especially careful.... but when I found the tail of [the riffle] dashing hard against the overhanging cliff at the foot [of the riffle] the temptation to give my passenger a chance to perpetuate on celluloid the back-thrown spiral was too strong to resist. Shurtliff closed his eyes ... when the cliff started to fold over but stuck gamely to his crank. My excursion under the cliff carried the *Grand* too far to the right to miss a shallow midstream boulder bar just being uncovered by the falling river, and she bumped-the-bumps all the way down, turning round and round in the surging current. Twice the gunwale was hove under as she struck a rock beam-on, but each time, luckily, there was just

---

33. Lewis R. Freeman, "Hell and High Water: The Grand Cañon's Roaring Rapids Vanquish the Explorers," *Sunset Magazine* (October 1924): 16–17.
34. According to an inscription in the back of La Rue's photographic notebook, Fox Movietone News photographer J. P. Shurtliff lived in Salt Lake City, Utah. Members of the expedition variously spelled his name Shurtleff, Shirtliff, and Shurtliff; we changed all spellings to the correct one. La Rue, photographic diary.

From left to right, Eugene La Rue, Mabel La Rue, Donald Dudley, J. P. Shurtliff, and Roland Burchard at the USGS building at Bright Angel Creek near Phantom Ranch. Frank Word is seated in front. Dudley was the USGS gaging station operator, Shurtliff the Fox Movietone News man. The image was probably taken on August 26 or 27.

Photographer unknown, Topography D 38, courtesy of the U.S. Geological Survey Photographic Library.

---

enough water to swing her clear. From the boating standpoint an upset in such a place would have been unpardonable; and yet that was very nearly what happened. Shurtliff was clawing red water out of his eyes with one hand and cranking with the other all the way down, and probably will never know unless he reads this how near he came to leaving his camera on the bottom of the Colorado. I have not yet seen the resultant picture, although I did read in a Chicago dispatch that a woman fainted on viewing it—doubtless from the illusion of spinning round and round.

FREEMAN, REPRISE: The afternoon was spent in segregating outfit for the next stage of the run. We shall be somewhat more heavily loaded than before—not just what one would prefer on a falling river. The latter now has but little over half the flow it had when we arrived at Suspension Bridge five days ago. KHJ read our latest report over the radio this evening. John also answered some questions I had asked him by letter and sent a personal message from my mother. Static was rather troublesome.

LINT: Emery, Blake, Edith, Dodge and I left the rim at 9:30 and made it to the foot of Bright Angel Trail by 10 o'clock.[35] Edith stayed there and the rest of us took the short cut, got Emery's boat, crossed the river, and went up to camp. Mrs. La Rue rode with me and she got a real kick out of riding the waves.

MABEL LA RUE TYPESCRIPT:[36] It had been decided at the hotel, the day before that I should go on with the USGS party for a short distance by boat. Passengers were not taken into consideration when the boats were constructed. The oarsman sits in the cockpit, and all others have to sit or lie on the covers of the hatches, and hang onto ropes that have been lashed around the boat for that purpose. Everyone wears a life preserver, and when I put on mine, it seemed so heavy that I felt sure I would sink like a ton of lead if the boat should be upset. It was a sleeveless jacket of cork, with a big collar that insisted on pushing off my hat. Of course I couldn't borrow a hat pin from the men so I had to hold my hat with one hand, and hang onto the rope with the other. The boatmen told me to turn on my face and I would find it the easiest way to travel. At first the water was very smooth, and I had just decided that all these preparations were unnecessary and that the [dangers] of the Colorado had been vastly overestimated, when we suddenly rounded a bend and I heard a roar ahead of me. I raised up on my elbows and took one look and then grabbed that rope with both hands. I felt that my hat was no longer of any importance, for perhaps I should never need it again. And about that time the engineer in the bow of the boat called back to me, "Does your husband know your favorite flower?" I had just time to answer, "Anything will do except water lilies," when we hit the first wave. The boat reminded me of a kicking bronco, for it did all it could to shake me off. It shot forward and then suddenly stood almost still. It dove down into the trough, until I felt sure we should strike the bottom and then it turned and shot over the top of the next wave, leaving the ground in an airplane will be no new sensation to me. Sometimes we struck the waves

---

35. Lint's stated time of departure and/or arrival must be in error—the group could not have covered the 7.9 miles from rim to river in $\frac{1}{2}$ hour, not is it likely it took $12\frac{1}{2}$ hours.

36. Mabel La Rue, undated typescript, box 10, folder 1, La Rue Collection. Mabel La Rue rode in the *Boulder* since La Rue thought Lint to be the best boatman. Otis Marston, interview with Mrs. E. C. LaRue at Pasadena, November 20, 1948, box 114, folder 4, Marston Collection. La Rue's sentiments about Lint's skill were shared by, at the very least, Moore and Birdseye.

sideways, and I all but slipped off completely. Occasionally the waves would kick on the top of the boat, and before we were through the first rapid, I was pretty well soaked with that muddy Colorado River water. But as the temperature was over 100°F I didn't mind it a bit. I almost enjoyed the next rapid and was sorry when we reached Pipe Creek where the rest of the party was waiting for us. When I sat up and looked around I found a bunch of kodaks pointed in my direction, and I'm sure it was taking an unfair advantage of me. I hadn't even thought of powdering my nose since we started and I know my hat was not at the proper angle. But in a few minutes we heard the cook call "Come and get it or I'll throw it out," and no time was wasted in gathering around the rock that served as a table. We had corned beef, baked beans, and hard round disks of bread that the men called phonograph records, with strawberry preserves, and it tasted like a feast for the gods. I hadn't complained, or even intimated that I was frightened, but the men refused to take me any further. They said the next rapids would be too rough, but I think they were afraid I would spoil the bad reputation of the Colorado River Rapids, so I mounted one of the pack mules that had brought supplies to the survey party and rode back to the hotel. This mule was named Napoleon and he deserved his name. Whenever we stopped to rest, he insisted on trying to pass the other mules and take the lead. The trail back to the hotel was so steep in places that I let go of the saddle horn and clutched Napoleon's ears to keep from sliding back to the river. But we arrived safely, and I am convinced that never again will I have such a thrilling experience as I had shooting the rapids of the Colorado River.

August 28, 1923 — 16,900 ft³/s

BLAKE: Bert Lizon[37] and Postmaster Kitner[38] arrived at camp before we were up this morning. They had left the top about 4 o'clock in order to be at the foot of the trail when we ran the rapid. We were delayed, however, until 2 o'clock on account of the surveying of a damsite at the mouth of Pipe Creek. Luzon and Kitner returned to Grand Canyon as soon as the boats were thru the rapid.

37. Hubert R. "Bert" Lauzon lived at the Grand Canyon from 1911–51, working as a cowboy, miner, tour guide, trail custodian, and constable. He married Edith Bass, daughter of pioneers William Wallace and Ada Bass. Among other adventures, he participated in the Kolb brothers' 1911–12 river trip. Bradford Cole, "Bert Lauzon's Grand Canyon," in *A Gathering of Grand Canyon Historians: Ideas, Arguments, and First-Person Accounts*, ed. Michael F. Anderson (Grand Canyon, AZ: Grand Canyon Natural History Association, 2005), 49–52.
38. James Kittner was the postmaster at the Grand Canyon in 1923.

Soon after lunch, which was eaten rather late, we came to Horn Creek Rapid, which we could tell was a real one by the sound long before we could see it. A big rock was near the surface at the head of the rapid and was in mid stream, making two channels to choose. Dodge and I landed on the south side of the river while the rest of the party landed on the north side. It was decided to run it on the south side, as by doing so some of the worst of the big combing waves could be avoided. Emery, Leigh and Freeman came thru in close succession, and were tossed about pretty roughly. Emery's boat was thrown completely on its side by a big wave but did not capsize. He was compelled to drop the oars and hold on to the side of the boat, while the oars were wrenched about by the current in such a manner that the gunwale of the boat was split and one of the heavy oar locks was bent. My boat was the only one taking a passenger thru. Dodge rode with me and had to hold on with all his strength to keep from being thrown from the boat. Near the foot of the rapid the boat was forced into a swift back current where three attempts were made before it could be forced into the current again. While in the eddy it looked several times as though a wreck was certain, as the boat was propelled by the boiling currents toward a projecting point of the cliff, but each time the backwash of the water kept the boat from smashing. While in the eddy the boat struck a rock with such force as to nearly knock Dodge from the deck of the boat. I then jumped for the shore with the rope, but slipped on a rock bruising my palm slightly and knocking the skin from my knee, but managed to hold onto the rock and the rope at the same time, though I was half in and half out of the water. After bailing some of the water out of the boat another attempt was made to get out of the back current, which was successful. Emery has a lame hip, caused by being thrown against the side of the boat.

KOLB, LETTER TO HIS WIFE:[39] Dear Blanche, Started to take Bert [Lauzon] on boat, but film ran out. Sending it up. Have developed and print made. Please write immediately and tell him to print all (except the positive I sent in.) on sepia or yellow brown as before. If Ellsworth wishes to loan me some of his film, please help him rewind it and load the retorts then Edith or you bring it in to Hermit. We all felt bad yesterday, dropping down so fast and getting in the extreme heat but salts and rest fixed me up this morning. It was good to have this vacation around here. All boatmen had a good swim in the pool last night. With love, Emery.

---

39. Emery Kolb to Blanche Kolb, n.d., box 5, folder 625, Kolb Collection.

STABLER: Surveyed Pipe Creek dam site, completing about 1 P.M. Birdseye and Burchard handled plane table and recorded and Dodge and I rodded. Still looks good after going over it in Survey. Saddle spillway elevation 2749. Boats bounded through Pipe Creek rapids to the applause of quite a party of tourists. Not much of a rapid if boat properly handled. Camped opposite mouth of Trinity Creek after boats ... ran Horn Creek Rapids. This is a vicious little fall of 9 feet in about 50 with quite heavy waves.

August 29, 1923 — 15,200 ft³/s

BLAKE: We were on the move early this morning. The river below the first riffle was still and deep, with few places to land. Then came a small rapid [Salt Creek Rapid] and more still water. The sound of roaring water could be heard tho no waves could be seen until within a hundred yards of Monument Creek, where the boats were stopped so the boatmen could inspect what lay before them [Granite Rapid]. The river here narrowed down and took an abrupt drop over huge boulders which caused the rapid. On the north the huge waves lashed the sheer cliff while the south half of the rapid was thickly strewn with boulders, many of which protruded above the surface of the water. It was readily seen that the big waves could not be avoided and after discussing the possibility of running thru a narrow channel between the rocks we discarded the idea. After lunch we crossed the river and viewed the north channel again. A heavy shower fell while we were deciding the best method of procedure, but we kept dry by getting under a projecting rock. When the run was started Emery went first, I following, then came Leigh and Freeman. The first boat went closer to the cliff than had been planned but came thru O.K., taking less water than any of us. The remaining run to Hermit Creek was made in about an hour where camp was made at the head of the rapid.

BIRDSEYE: Col. Crosby,[40] Superintendent of Grand Canyon National Park, had made up a large party to come down to see us run Hermit Creek Rapids, and the Col. with a Park ranger, rode in on mules after dark. Shurtliff of the Fox News and Roger Birdseye also came in, as did Mrs. Kolb and daughter Edith. During the evening Shurtliff of the Fox News and La Rue took flash light movies of our radio set in action. The flare was too close and the smoke fumes nearly suffocated Stabler and me. Freeman also took still pictures with his large camera.

---

40. Walter W. Crosby was Grand Canyon National Park's second permanent (not acting) superintendent; Anderson, *Living at the Edge*, 110.

Downstream view of Hermit Rapid showing one of the extremely rocky debris fans that create rapids in Grand Canyon.

E. C. La Rue 458, courtesy of the U.S. Geological Survey Photograph Library.

LINT: Emery quit tonight on account of the way that La Rue and Freeman have been knifing him in the back. They are underhanded about it and those of us that know how things are don't blame him a bit. It would be far better to keep a good man and let the three drones (as they are) go out.

FREEMAN, DIARY: The reason [for Kolb's resignation] is not quite clear, but will doubtless be in a day or two.

LA RUE QUOTE:[41] Let Kolb stay home. We have good boatmen.

BIRDSEYE, REPRISE: There had been constant petty friction between Kolb and the crew of the *Grand*—La Rue, Moore and Freeman—most of it due to Kolb's extreme jealousy of anyone who writes about or takes pictures of any boat trip through the canyon. He considers these activities to be his special prerogative. It is quite true that he and his brother Ellsworth have had some rather hard knocks from the Harvey Company and the Santa Fe Railroad, the major concessionaires at South Rim, and have had quite a fight for existence as scenic photographers in the Park. Nevertheless the Kolb Brothers failed to realize that in this progressive age they must survive by competition and not by adopting a "dog in the manger" attitude. La Rue and Kolb mix like oil and water and have had a

---

41. Mabel La Rue, interview with Otis Marston, November 20, 1948, box 114 folder 4, Marston Collection.

little friction, but not enough to force my notice. Kolb has picked on Freeman incessantly and the latter has shown a fine spirit by not saying a word. Freeman brags a little about his past explorations and his writings and all this seems to have grated on Kolb's sensitive nature. The moving picture venture of the Fox News and also La Rue's operation with the little Sept camera have also displeased Kolb who at first had asked for exclusive moving picture rights. He has feared that the Harvey Company would get the pictures and run them as a free show at the El Tovar Hotel and thus ruin his business which includes two moving picture shows a day at his studio. There seemed nothing to do but to make the best of a bad situation. Several bad rapids lay ahead, and none of the party had ever seen them except Kolb. So I told him to go. I then arranged with Dodge to take the fourth boat and decided not to have a head boatman, but to try to work out harmonious action from the four boatmen. Expected to rely largely on Lint and Blake who have had more experience in rapids of this kind than Freeman. Lint handles his boat much better than anyone in the party, including Kolb, but he is too young for the responsibility of head boatman. Then, too, Kolb has evidently soured him on Freeman, so there would continue to be friction. In the meantime, Mrs. Kolb and Edith came down and were evidently very much shocked by the way Emery had acted. At about 8:30 P.M. Emery came to me and begged to be taken back. I told him it was too late. He then asked me to try him as far as Bass Trail and said that Blake, Lint and Dodge joined him in the request. Dodge said he wanted to have a boat but would give up the chance if Emery would stay. Then consulted with Freeman who showed a fine spirit and said "by all means." Then talked it over with Stabler who also advised me to give him another trial. I then told Emery that we would try again and he agreed to act decently and make no criticism or remarks about anyone. When Emery gave my decision to Mrs. Kolb and Edith the latter fainted from the reaction and we had quite a time, using both the ammonia and the "snake bite" bottle. Edith stayed in camp that night while her mother went up to Hermit Cabin, 1½ miles up the creek. Edith seemed OK in the morning but rather weak.

DODGE, LETTERS: [Dodge's view of the situation was revealed by comments made in a series of letters written to river historian Otis "Dock" Marston in 1947. Had Kolb left, Dodge noted, "I don't think any one would have rated head boatman ... Lint & Blake were'nt too hot and were sort of trouble makers and quite young.

I would have advised Birdseye to hold a pow-wow at each bad place and let the majority rule."[42] Dodge also described his reaction when Birdseye informed him of Kolb's request to continue with the expedition, "Frank, Emery wants to come back but it's up to you; I've given his boat to you." I laughed. I was working for the Gov't and Birdseye. What the hell did I care as long as the trip was a success so said, "Take him if you want him and remember I'm only a hired hand. Better take him otherwise we'll be a man short. Incidentally, Birdseye was the finest boss I ever had."[43] As for Kolb himself, Dodge wrote, "Am afraid Emery Kolb would'nt [sic] like me as well after reading my biography as before, You know, Emery has a 2x4 mind, a jealous character, and a view point hard to understand. He, La Rue and Freeman were in constant turmoil all due to Emery's imagination over who was to take the pictures ... Every one who runs the Canyon is also a S. of a B. Jealous of all who go thru. Yet, I always think of him as a friend but a narrow minded friend.[44]]

August 30, 1923 — 15,400 ft³/s

BIRDSEYE: The cook was on a rampage because of so many visitors and has announced his intention of going out at Bass Trail—our next stop—so have sent out for a new cook to join us at that point. The breed of cooks seems to be composed of nothing but cranks and grouches. Word is a good cook, although slow, but seems to think the whole party is to suit his convenience and growls on the slightest provocation. We have all treated him far better than he was ever treated before but it is the same old thing—"you can't make a silk purse out of a sow's ear." Col. Crosby and party of about 30, including Mrs. Crosby and several ladies, several Santa Fe Railroad officials, Harvey Company officials, etc., came down from Hermit cabin to see the boats run Hermit Rapids, a very bad and rough one. Emery Kolb seemed to be in good humor, in spite of the Fox News and the Santa Fe movie operators who were there, and posed for a close up with the other boatman. After careful preparations, both by the boatman and the camera operators, Kolb and Lint started in with the *Marble* and the *Boulder*. The light was good and I think the Fox News will have some good pictures. The Fox News will give the Survey a

---

42. Frank B. Dodge to Otis Marston, December 27, 1947, box 51, folder 18, Marston Collection.
43. Frank B. Dodge to Otis Marston, November 27, 1947, box 51, folder 18, Marston Collection.
44. Frank B. Dodge to Otis Marston, November 4, 1947, box 51, folder 18, Marston Collection.

reel of what Mr. Shurtliff has taken of our operations and the pictures will doubtless be shown in many theaters over the country.[45] The boats tossed about like corks on the huge waves, many of which broke completely over the boats, which nevertheless went through in fine shape. This rapid is #15 and has a fall of 15 feet. After giving the camera men time to change stations and film, Blake and Freeman started through with the *Glen* and *Grand*. Blake carried Dodge lying prone on the aft hatch—the boats are run stern first[46]—and as each wave submerged him he raised up with a grin and wiped the muddy water from his bald pate. The *Glen* and *Grand* also went through in fine shape. The camera men and others watching from the bank started back to Hermit Camp and the rest of us had a hard climb over the low cliffs to where the boats had landed below the rapid.

STABLER: Cook wanted to ride [through Hermit Rapid] and had a grouch all day because Emery didn't take him.

LINT: Freeman followed [Blake] immediately afterward and shipped quite a bit of water. He stuck close to the celebrity and the women all morning and didn't even take time to look the rapid over closely. Worked down to Boucher Rapids and ate lunch. This rapid, No. 16, has 12 feet fall and has a nest of boulders on the south side and huge curling waves on the north side. Ran it on the north side in the usual order, all of us taking on considerable water. Camped on the north side of the river just above the mouth of Crystal Creek.

FREEMAN: This creek [Crystal]—at present as clear as thick red mud can make it—has thrown out a huge boulder delta, below which is a long shallow riffle of no considerable precipitancy.[47] Moore and I celebrated our departure from the last contact with civilization by clipping each other's hair. There was bad static interference with the radio. Stabler claims to have heard fragments of a Bedtime Story about Moses in the Bulrushes, and Moore avers he heard Italy threaten to declare war on Greece, but without learning why.[48]

---

45. The notion that film footage of Grand Canyon would be shown in theatres across the United States was exactly why Emery Kolb threatened to leave the expedition. This suggests that Kolb's concern that USGS was insensitive to his business may have been valid.
46. Nathaniel Galloway developed the stern-first technique, which remains the choice of most Grand Canyon boatmen because they "face their danger."
47. Crystal Rapid would become one of the most challenging in Grand Canyon in 1966, when a debris flow brought on by prolonged winter rainfall threw a much larger "delta" into the Colorado River; M. E. Cooley, B. N. Aldridge, and R. C. Euler, *Effects of the Catastrophic Flood of December, 1966, North Rim Area, Eastern Grand Canyon, Arizona* (U.S. Geological Survey Professional Paper 980, 1977).
48. Members of an Italian commission assigned to establish the Greek-Albanian border were massacred by a group of Greeks, leading to swift retaliation by Italy and stirring

Emery Kolb rows the *Marble* through Hermit Creek Rapid.

Lewis Freeman photograph, Grand Canyon 313, courtesy of the U.S. Geological Survey Photographic Library.

August 31, 1923 — 23,600 ft³/s

BLAKE: Some time during the night we were awakened by thunder and lightening. Rain began to fall and the wind started to blow. Emery, Leigh and I had made our beds close together, so that one tent fly would do to cover all of us, but our combined efforts were hardly equal to the task of holding our shelter over us as the wind had risen to a gale and the rain was coming down in torrents. Soon we heard water running down the slopes and began to get uneasy. In fact a small muddy stream was starting to flow between our beds and was rising every minute. Suddenly a larger stream broke over the sand bank near our heads, whereupon we each gathered everything we could get our hands on and climbed to higher ground. Leigh went back and tried to pull his small bed tarp out of the mud which had covered it, but was unsuccessful. He did return, however, with his own and my trousers, which he had salvaged from the muddy stream, and threw them down saying, "there goes

---

fears of another world war. "Commission of Italy Massacred," *Los Angeles Times*, August 29, 1923 (accessed February 28, 2006); "Kill Greeks at Corfu," *Los Angeles Times*, September 1, 1923 (accessed February 28, 2006).

our watches," meaning that our time pieces were probably ruined by the muddy water. Emery and I crawled between our blankets and made ourselves comparatively comfortable, while Leigh went to look after the boats. He returned presently reporting that ... the tent fly was down, that the cooking utensils were partly covered by the sand which had been washed thru the kitchen, and that the river was rising. His bed was not in shape to use so we all went in search of drier quarters, finding an overhanging shelf which would partly shelter two of the beds. Leigh took his blanket and lay down under the shelf while the Colonel, Emery and I used our beds with the mattresses inflated. By morning the rain had nearly ceased and soon everyone was stirring about. Leigh's life jacket had been blown into the river so was lost. After things had been straightened up a bit and the beds put out to dry a hearty breakfast was eaten.

LINT: This morning we went up and dug my tarp and shoes out of the sand. They were buried about 8 inches deep. Several of the grub boxes were half buried and we had quite a time finding one of the Dutch oven lids. The Colonel's bed was in a small lake and La Rue swore that his bed was floating but "I don't know, it may be so—but it sounds so dog-gone queer." It took until eleven o'clock to get the beds dried out. The water-proof box that the alidade was in was washed about 20 feet down the bar and was full of water but no damage done.

BIRDSEYE: Reached the mouth of Tuna Creek at mile 37.2 and found it running about ½ second foot of good clear water. Tuna Creek Rapids (#18) start at the mouth of the creek and have a total fall of 13½ feet, although really made up of two rapids, the upper 9½ feet and the lower 4 feet. All boats except the *Grand* pulled in on the left bank between the two parts of the rapid and stopped for lunch. The *Grand* was carried by and the crew had to resort to salmon for lunch, these being about the only articles of food aboard that boat. There are three large boulders in the channel at the lower end of this rapid [now informally known as Nixon Rock] and these afford dangerous passage, not so much on account of hitting the rocks as being drawn into the large whirlpools below each rock. All boats, however, made the passage in fine shape. One cannot walk around this rapid. Ran a small riffle with a 3-foot fall at mile 38.5 and reached the mouth of Sapphire Canyon at mile 39.2. There is not much evidence of a side canyon at the water's edge, so can not pass on the merits of the name, nor those other side canyons in the near vicinity called Agate, Turquoise and Ruby. Sapphire Canyon Rapids (#19) start at the mouth of the canyon and have a 7-foot fall. There are no rocks near the surface at medium stage of water such as that on which we are riding, but the waves run high and all boats took considerable water although I did not get wet at all. One can not walk around this rapid so we all rode. Ran a small riffle at the mouth of Turquoise Canyon at mile 39.9 and another small one at mile 40.7, both having a fall of 4 feet, with short choppy waves. In low water stage these might not be noticeable, but both are shown as rapids on the Shinumo sheet. Here let me record that I have never followed any maps which expressed the features with such remarkable accuracy as the Vishnu, Bright Angel and Shinumo [15'] topographic maps. We find slight discrepancies in minor detail but Matthes and Evans [Francois E. Matthes and Richard T. Evans, USGS geologists] surely deserve great credit for the work they did.

BLAKE, REPRISE: A heavy rain fell just after lunch, but as soon as it ceased we started work again. Scarcely enough soil was seen during the afternoon run to set foot on. Most of the landings were made on projecting shelves of rock or the sloping base of the cliffs, where pot holes worn by the action of the water, made foot and hand holds. In several instances I would let Dodge off on a point of rock, then drop the boat behind into the backwater where it would ride easier, and when ready to move on row close enough so he could step or jump onto the deck.[49] After making one difficult landing, Dodge having jumped shore to a slick granite boulder, I followed, and taking the rope, let the boat have some slack, but when it came to the end of the slack line I found I was unable to hold it, so was pulled into the river. I easily climbed into the boat and soon had it in the quiet water behind the point.

STABLER: Enjoyed riding the rapids today. It is great sport and not dangerous except perhaps in the rarely dangerous ones. Got well drenched with spray.

FREEMAN: We are camping tonight on a sand-veneered terrace about five miles above the Bass Trail. It has rained twice today and everyone is busy building stone platforms for their beds to guard against the catastrophic washouts of last night.

---

49. The film footage of the 1923 expedition shows that some survey points were taken from the boat while moored against a vertical cliff.

September 1, 1923 — 20,400 ft³/s

BLAKE: About 3 o'clock this morning the cook woke us saying that the river had dropped and that the boats were standing on end. We got up and found that all of the boats were half out of the water and there was danger of the *Glen* and *Marble* breaking their ropes if the river dropped farther, we therefore pushed them into the water but left the other two boats stay as they were, as they were resting on shelves of sand sufficiently wide to keep them from tipping backward as the river receded. This morning we found that the river had dropped four and a half feet during the night.

BIRDSEYE: Ran a 4-foot riffle at mile 41.8 and surveyed dam site #9 at mile 41.9 [above Ruby Rapid]. This took all the forenoon and did not impress me as a suitable dam site, as neither walls nor spillway location were good. However, it was at the proper elevation for a low-power dam that would not flood Phantom Ranch in Bright Angel Creek so we made the survey.

FREEMAN: This forenoon I repeated my experience of yesterday in running dry through a fairly rough riffle with all hands, and then slopping in several inches in an innocuous looking drop not far below. The rapid below Serpentine Creek, marked on the map as half a mile long—looked troublesome with its upper end choked with boulders and a heavy surge of water against the sheer right-hand cliff. The latter seemed avoidable, however, by putting in well to the left and pulling away from the wall. (Stabler inclines to the opinion that this is the rapid characterized by Stanton as "the worst and most unmanageable on the river," and where he lost his *Marie* by trying to run her through empty without boatmen or line).[50]

LINT: Worked on down about a mile to the Ruby Creek Rapid (No. 20). This rapid ran into the wall on the north side so we started in the center of the V and pulled to the south bank. Everybody rode through but got a good wetting. 8½ feet fall. Just below this was a large riffle with 4 feet fall that had some real waves in it. At 2:30 we came to the Serpentine Creek Rapid and we spent about half an hour looking it over. It is No. 21 and has 10½ feet fall. All of the water runs into a pile of boulders on the north side making a nasty lot of waves and boils. We entered it on the south side in a small V and passed close to a huge submerged boulder and then pulled

---

50. As most novice Grand Canyon boaters are, Stabler is "lost in the Jewels" even if Birdseye knew exactly where he was. The incident Freeman discusses concerning Stanton and the destruction of the *Marie* occurred in Horn Creek Rapid about fifteen miles upstream; Smith and Crampton, *Colorado River Survey*, 177–78.

through some large waves and into the back current. Everybody walked with the exception of Dodge who rode through with Blake. Freeman is getting so good that he don't even have to look the rapids over. He is just too lazy to do any walking except when there are strangers around so that he can show his noble strut. (Some walk he's got.)

STABLER: Camped just above Bass Crossing.[51] Put cook outfit on a point of rocks but river came up about three feet and made an island of it so it had to be moved. Likewise I moved my bed from its original location though the morning showed I could have left it by a narrow margin.

September 2, 1923 — 27,500 ft³/s

FREEMAN: The river began rising rapidly just before dark last night ... Everything was moved over to the right bank before the channel became too swift to wade, and the boats were brought in and pulled up as high as possible. The rise to midnight was about three feet, after which it only came up a few inches. The fact that it is holding almost steady at its high mark this morning would seem to indicate that the flood is from the Little Colorado or San Juan, rather than local. Moore, in great glee, brought in a mighty hunk of brown rock this morning, which he says contains "calcareous algae," probably the oldest form of life. They seem to have been dead a long time, however, and their tomb is heavy enough to give the *Grand* a sharp list to port. I am trying to induce Moore to consume enough salmon and sardine from our cargo to offset his augmenting load of rock, but not with marked success.

BIRDSEYE: The pack train arrived at the foot of Bass Trail at 12 PM bringing heavy supplies of canned goods, etc. to last until Diamond Creek—planning to have only light supplies such as mail, sugar, flour, etc., brought in at Havasu Creek, when the last 6 miles will be man-packed. The heavy canned goods taxed the capacity of the already overloaded boats, but were finally stowed away.

FREEMAN, REPRISE: Roger Birdseye brings word of a violent series of cloudbursts in the region of El Tovar, doing great damage to roads and trains. This would be the same cataclysmic disturbance that tried to put us into the river our first night below Hermit Creek. Roger was on time as usual with the packtrain. His work in

---

51. The Bass crossing and trail was named for William Wallace Bass, who, with considerable assistance from his wife, Ada, ran a tourist business in the Grand Canyon. Bass also dabbled in mining. Anderson, *Living at the Edge*, 42–52.

this connection has been invaluable. I am inclined to think he has worked harder most of the time than anyone on the river party.

KOLB, LETTER:[52] Dear Blanche & Edith. The mules got here a little earlier than we expected, so I can only get out a short note. We have had no trouble. Some laffy rapids though. Big rain got us up 2 am one night six inches sand carried under our beds. I don't know if Frank [Word] is going out or not. River is running high but makes less rocks. Please have Edith write and tell that Iowa College that I am with the USGS and cannot say definitely when I will be east with my lecture but expect to be there possibly Jan, Feb, & March. Also with a new lecture. "Hazards of preliminary work of damming the Colo. River. One lecture $150.00 $100 a lecture if more than one is arranged for. Give details of my new trip with USGS engineers surveying damsites. With much love, Dady. [P.S.] 4 P.M. Now that I have a little more time I will write some more. I wish you would open the package of movies and wind them over a reel (that is the positives) and send me some word at Diamond Creek about how they are. Just tell me how much negative and how much positive they printed then I can draw my conclusions. Well here is morning again so I must close and say good by for another month.

BIRDSEYE, REPRISE: The pack train left on the return trip at 2 PM and the river traverse was continued. Ran Bass Canyon Rapids (#22) at mile 45.9. This rapid is small and short, but rough, having a fall of only 3 feet. Passed under the upper Bass Cable Ferry at mile 46.2. This cable is 900 feet down stream from the point shown on the Shinumo [15'] sheet. It seems to be a well constructed affair with heavy support cable and 2 pulling cables operated by a windlass on top of the car (man power). The cable is capable of ferrying an animal and also carrying a man on top of the car to operate the windlass. Rex Beach[53] lost a horse in crossing here, the animal being frightened in the car and hanging himself on the tie ropes. The cable is about 50 feet above the water and 300 feet long.[54]

---

52. Emery Kolb to Blanche and Edith Kolb, September 2, 1923, box 5, folder 624, Kolb Collection.
53. Rex Beach was a prolific adventure novel writer; many of his books were also made into films.
54. The National Park Service, believing both cableways near Bass Canyon to be a hazard to aircraft, removed them in 1968; Kim Crumbo, *A River Runner's Guide to the History of the Grand Canyon* (Boulder, CO: Johnson Books, 1994), 37. Curiously, no one mentions seeing the Ross Wheeler just downstream from the mouth of Bass Canyon. That boat, built by Bert Loper, was abandoned on the left bank by Charles Russell in 1915; Brian, *River to Rim*, 86–87.

BLAKE: We worked down the river about a mile to Shinumo Creek, which is a nice stream of cool water, tho somewhat red from the recent floods. There was no wood at the mouth of the creek so Emery semaphored to Leigh, who had landed a couple thousand feet above with the surveyors, to bring some wood. Emery and I then went up the creek to a waterfall and seeing some driftwood above the falls we went back into the cave which we found to the right of the fall, and found a way to get up above. I went up, struggling through a narrow drift-chocked opening, to where I could loosen some of the drift and throw it down. Leigh brought a load of mesquite wood which with what we had gotten, was ample for our needs. The river began falling soon after we made camp. We had to loosen the boats before going to bed, and during the night it was found that the *Glen* was resting upon two rocks and was completely out of water, while the other boats were tilted at a steep angle from the prow to the stern.

September 3, 1923 — 18,800 ft³/s

BIRDSEYE: Lay idle all day, calling Monday Sunday for a change. We can take liberties with the calendar when in the Colorado Gorge.

STABLER: Camped at mouth of Shinumo Creek all day. Before breakfast I took a bath in a pool in the creek and did my week's washing—not very extensive. Then walked down the river 50 yards until a sheer cliff at the water's edge stopped me. After breakfast Moore and I climbed the point east of the creek and took a look at the saddle spillway of the suggested dam site. It is faulted in three places and the rocks are rather badly shattered. It is not suitable for a spillway and would probably leak if water were backed up against it. Hence the dam site goes by the board.

BLAKE: After breakfast we took pictures of the *Glen* which was resting two feet above the water with only the prow and one corner of the stern being supported by the rocks where she had settled when the river fell. Most of us went bathing at the falls in the creek. Leigh, Emery and I went above the falls and jumped off into the pool below. Going back to camp, Emery, Leigh and I took the *Glen* and rowed it up the river a hundred yards where we found some shade under a high boulder. After lunch we took turns trimming each others hair, then we went back to the shade of the big boulder, Emery and I swimming back. We spent the afternoon writing and sleeping.

FREEMAN: La Rue finds the creek is flowing 16 second-feet—the largest we have passed with the exception of Bright Angel, which had a flow of 36 second-feet a week ago.

September 4, 1923 — 15,100 ft³/s

LINT: Ran the Shinumo Rapid (No. 23) the first thing this morning. Large waves but no rocks. 7 feet fall. Emery broke one of his spare oars in this rapid. He had it lashed on the aft deck and a large wave caught the blade and snapped it off. About three quarters of a mile below Shinumo Rapid is No. 24 with a fall of 10 feet. It is a long rapid in a bend of the river but is not a bad one.

BIRDSEYE: ... came to the lower Bass Cable Crossing at mile 48.8. This is rather a flimsy affair and is designed to carry one man only—the car being small and the supporting cable only about ¼ inch in diameter. The cable is about 175 feet above the water and about 600 feet long.[55]

STABLER: Broke camp after breakfast and carried the line down to Hakatai Canyon where a dam site survey was completed about 2 P.M. (Excellent site). We reached Walthenberg[56] [*sic*] Canyon about 3.30 P.M. and by 4.30 were ready to move on but as camp sites seemed likely to be scarce we made camp there at the head of a sharp rapid that will give us a wetting in the morning. There is a pretty fair dam site here at Walthenberg Canyon. The right abutment is the best I have yet seen, but the left wall has a rather flat slope. The spill into Walthenberg Canyon would be OK. The rapid appears to be formed partly at least by three pegmatite dikes that jut into and partly across the river.

BLAKE: This rapid is where Emery smashed his boat on a former trip, and where Ed. Kolb's boat capsized and rolled over and over by the backwash of a big rock, Ed. being carried a half mile down stream before he could land. They spent Christmas day repairing their broken boat.

---

55. The lower Bass Cable was also removed by the National Park Service. The cable was left in the river and can be seen on the left bank and high on the right bank.
56. This creek and rapid is named for John Waltenburg, who worked for Bass just upstream; Crumbo, *A River Runner's Guide*, 38–39. Unfortunately, either the USGS or the U.S. Board on Geographic Names chose to spell the man's name Walthenberg, and this name persists on USGS maps and in some river guides.

6

Of Flips and Floods

Bass Canyon to Diamond Creek

In early September, the 1923 USGS expedition found itself out of the Inner Gorge and away from seemingly continuous rapids, but all was not well within the crew. Although Birdseye had placated Kolb, who chose to remain with the expedition after his mutiny at Hermit Rapid, Birdseye faced an undercurrent of bickering. Word, who had the ready excuse of eye problems, would leave the trip but later blame Kolb's attitude for at least part of his decision. La Rue continued to irritate, Freeman failed to pull his weight, and the others chose sides. While many diarists noted conflict among the crew, Freeman elected not to do so. How much of this Birdseye chose to deal with or ignore is revealed somewhat in his diaries and actions, which continued to hold the trip together.

While the crew was in interpersonal purgatory, the scientific mission proceeded with few hitches. The weather continued to vex the crew, who alternately experienced extreme heat, wet cold, and blowing sands. The river rose and fell daily, resulting in stranded boats; modern river runners associate this phenomenon with the diurnal releases from the power plant of Glen Canyon Dam. Of the few large rapids facing the expedition before their next resupply at the mouth of Havasu Creek, one caught the full attention of the head boatman. This gave the until-now unsung rodman a chance to prove his aquatic athleticism, and get a pay increase as a reward. The crew sent their radio out for repair at Havasu Creek, only to miss the dire warnings of an impending flood, one that they would survive but which would leave the outside world wondering.

Eugene C. La Rue measures flow in Deer Creek below Deer Creek Falls, the prominent waterfall at river mile 136 in Grand Canyon.

Photographer unknown, La Rue 545, courtesy of the U.S. Geological Survey Photograph Library.

September 5, 1923 — 13,700 ft³/s

MOORE, LETTER:[1] We have had a most interesting trip from Bass, most of the way traveling in a deep narrow gorge carved in granite. We have absorbed so much canyon scenic grandeur that we've almost reached the saturation point.

BIRDSEYE: Number 26 rapids [Waltenberg Rapid] looked worse in the morning and we found it had a fall of 13½ feet with a strong current below, so all walked around as far as a projecting cliff on the left bank. Dodge swam around this cliff with a rope and stood on a rock below ready to throw the rope to any of the boatmen who needed assistance. As usual, Leigh Lint made the best run with the *Boulder* and pulled in just above the cliff. Lint ferried all passengers past the cliff and we considered ourselves well past a bad rapid which every other party had either lined or portaged. Reached Royal Arch Creek, coming in from the left, at mile 54.6. The lower end of this side canyon is known as Elves Chasm and the boys in the boats which stopped there say it is the prettiest spot on the river not excepting Vasey's Paradise ... there was some disappointment when Burchard and I sailed by in the *Boulder*. However Stabler was counting on an extra day at Tapeats Creek and as it was only 3 P.M. he did not give us the camp site signal. He had picked out another place four miles below which looked as though we might find fresh water—but we had no such luck. During the afternoon we discovered the collimation of the alidade to be in bad adjustment and decided that it had been caused in cleaning the instrument after the storm of the night of Aug. 30. No change in this adjustment on many previous tests had made us over-confident and we had not examined it since the storm. I determined the error per 100 feet of distance and applied a correction at each set-up for excess of back or fore sights and found an accumulative error of -2.0 feet. Decided to make a tentative correction of this amount and give it additional consideration when we checked on Evans's point at the mouth of Havasu Creek. Camped on the left bank at the beginning of Conquistador Aisle at mile 58.2, after a day's run of 7.8 miles [probably near Blacktail Canyon].

BLAKE: We spent considerable time looking over the rapid this morning. Leigh struck an oar on a rock in mid stream as he came through. Emery followed, making a good run. I then came thru, fol-

---

1. "Death Dodged by River Party," *Los Angeles Times*, October 5, 1923 (accessed March 7, 2006). Moore wrote a letter to his mother, Mrs. B.H. Moore, which was quoted in the *Los Angeles Times*.

lowed by Freeman. Dodge, wishing to try out his rope, threw it to me and the knot end of the rope caught in a line amid ships, drawing the boat sideways and causing it to take water until I managed to loosen it. I had no difficulty in landing but was in such position when I resumed rowing that the boat struck a boulder, tho doing no damage.

STABLER: About three o'clock we came to Elves Chasm. This is a beautiful spot at the mouth of Royal Arch Creek. A perennial stream of excellent water, grass, flowers, trees, ferns, and aquatic vegetation, with beautiful water falls all joined to make the place attractive. An overhanging cliff of Tapeats sandstone made an excellent sheltered camp site. There was the remains of an old camp fire there, and the names of R. L. Elliott, Norman Oliver, and Steward (?) with dates of May and June, 1905, were written on the rock in charcoal—also the letters USGS—perhaps the only ones besides ourselves who ever saw this beauty spot for none of the river parties have mentioned it.[2] It was too early for camp so we passed on 4 miles and camped on a huge sand bar at the beginning of Conquistador Aisle.

FREEMAN: The river has continued falling at a surprising rate and a good many rocks are coming up in the rapids to restrict the channels. I hit two in getting into a little riffle a mile and a half above camp, running squarely over the first and hitting the second a sideswipe with the stern as we swung. The clear channel on the right was plain enough, but I could not make the boat move over to it fast enough when it opened up. If the river continues to fall at the present rate bumps are likely to become rather common.

September 6, 1923 — 12,900 ft^3/s

STABLER: After breakfast we crossed the river to the creek coming in from the right and got fresh water. This little canyon is cut through Tapeats sandstone though the creek bed for a few hundred yards at least is in granite. The walls overhang and in places overlap so as to obscure the sky. Emery took a 3A and a movie.[3] We then

---

2. Nellie C. Carico, while doing topographic history work for the USGS in the mid-1960s, made a note in Birdseye's diary that the men were Norman Oliver, Herman R. Elliott, and John T. Stewart; Elliott, and presumably the others, was with the Topographic Branch of the USGS. Clyde Eddy and his crew saw this inscription in 1927; Clyde Eddy, *Down the World's Most Dangerous River* (New York: Frederick A. Stokes Company, 1929), 202. The topographers were collecting data for the Shinumo Quadrangle map; Edward Morehouse Douglas, Richard T. Evans, H. L. Baldwin, and John T. Stewart, "Arizona (Coconino County), Shinumo quadrangle" (Washington D.C.: U.S. Geological Survey, 1908).
3. Stabler apparently is discussing the first visit to Blacktail Canyon, now a common stop for river trips through Grand Canyon.

Upstream view of 128 Mile Rapid on September 6 at 5 p.m. The man seated in front is probably Leigh Lint or Elwyn Blake.

E. C. La Rue 520, courtesy of the U.S. Geological Survey Photograph Library.

---

had a beautiful ride through Conquistador Aisle about 4 miles and lunched in the shade of rocks where the river turns northeastward.

BLAKE: Only three rapids [Forster, Fossil, and 128 Mile rapids] were run today, but the river has dropped fast, causing many riffles between which there is still water. The only rapid which looked serious was the last one run, it having about seven feet of fall in the first fifty feet, which caused very rough twisting waves for another hundred and fifty feet. Freeman's boat went thru more jerks and twists while coming thru than I have seen any boat make in such small waves (not over seven feet high). Leigh's boat did not take as much water as the *Grand*, but was whirled around several times in the whirlpool at the foot of the rapid. I did not see Emery, who had my boat, come thru, but he said he took quite a bit of water. I cut to the right of the V with the *Marble*, and only took a couple of splashes. We did not reach camp until about 6:30, and seeing a nice pail of clear water conveniently situated, took a big mouthful, but not a swallow, for it was very salty, and had been placed there by the cook for everyone's benefit, but especially for mine, as I had played a similar joke on him up in Marble Canyon. The flies are so vicious tonight that most of the fellows have put on their shirts.[4] The last two days have been very hot and the nights are sultry until the rocks cool off.

BIRDSEYE: Camped at Specter Chasm at mile 67.2 after our longest day's run of 9.0 miles. Stabler had picked this point with the expectation of finding good water at a "chasm" but we were all disappointed. A small trickle of alkaline water was our luck and the place lived up to its ghostlike name. We were all too tired to put up the radio and there was no suitable location.

STABLER, REPRISE: I geologized a few hundred yards up the creek [in Specter Chasm] for Moore's benefit in case he did not make it. Emery, Frank and I are now (5 P.M.) ahead of the rest with the cook boat and have gotten camp started. Our arrival here evidently startled a mountain sheep as he left very fresh tracks at the mouth of the creek.

September 7, 1923 — 12,200 ft³/s

STABLER: We started out with the Specter Chasm dam sites (nos. 13 and 14) surveying about a mile of river below Specter Chasm in detail. Two excellent sections with spillway facilities by tunnel are available. It would be possible to put a dam here with water elevation 2000 to back to Bright Angel Creek elevation 2435, 41 miles upstream.

FREEMAN: We ran the rapid below Specter Chasm [Specter Rapid] immediately on pushing off. In avoiding the high rollers against the cliff, Lint took the *Boulder* through a breaking wave over a rock on the left of the V, completely submerging Burchard, on the stern. Kolb ran the V and had a good passage. I got over a bit too far to the right, and so hit the tops of the high places all the way down, shipping some water and impartially wetting both my passengers and, of course, myself.

BLAKE: The second rapid encountered today [Bedrock Rapid] was one of the worst, from one standpoint, that we have encountered. The whole force of the river was directed against the outcrop of granite, the right hand channel being the only reasonable course, while the left channel carried the most water and drew all

---

4. One of the misconceptions prevalent among post-dam users is that insect problems are associated with heavy, post-dam river use. Most don't realize that problem insects were perhaps more common to pre-dam river runners; Webb, Melis, and Valdez, *Observations of Environmental Change*, 24. In this chapter, members of the 1923 USGS expedition discuss mosquitoes, flies, gnats, scorpions, and tarantulas.

floating matter its way, and if a boat should be drawn to the left it meant a certain smash up on a rock lower down, and also meant being dragged into a back current from which there was seemingly no way out. I took Dodge, and getting a good start, pulled prow first into the interference waves on the right near the head of the V and by pulling hard missed the main rock by over twenty feet, but drifted over a submerged rock in the quiet water at the foot of the rapid.[5] I then took Dodge a couple thousand feet below where he held the rod for a turn. Emery then came thru and reported having come within two feet of the big rock, which statement was refuted by the cook, who claimed that the boat only missed by one foot. I took my camera back up the river to get some pictures, walking nearly a half mile, partly over blistering hot rocks bare-footed. I then came back to the boat, or rather Emery's boat, as he had taken my boat and gone on ahead to make camp. Dodge wished to go back and bathe in some clear pools I had found, so I went with him. After bathing we climbed upon a cliff which overlooked the rapid and waited for the *Boulder* and *Grand* to come thru when the surveyors returned from the survey of a side canyon. Leigh pulled hard but did not get the start desired and came so close to the rock that Burchard, who was riding in a prone position, upon the stern, slapped the rock with his hat.

MOORE, LETTER:[6] Our boat with all of us on board had about the narrowest squeak yet at one place below Bass. In running one rapid, with the walls sheer on both sides, the full force of the current piled up against a little rocky islet in midstream, split, and with a big swirl and eddy below, filled with rocks, went swiftly on [Bedrock Rapid]. To go to the left meant almost certain disaster and to skim too close to the right was equally dangerous. Our boat, the biggest and heaviest, was a little more than Freeman, our boatman, could manage to pull to the right. We hit the rock almost dead center. With a bang on our stern we teetered a second, then, swung right and shot down, just skinning by rocks that would have ripped the boat wide open.

BIRDSEYE: Finished the survey at 5 P.M. and had expected to camp apart from the cook boat which we had sent on, presumably to Tapeats Creek. Ran #31 rapids at mile 68.5 [Bedrock Rapid].

---

5. This rock still claims unwary boaters, who are so relieved to be by The Bedrock that they fail to notice the tell-tale signs of swirling water around this rock at most present-day river levels.

6. "Death Dodged by River Party," *Los Angeles Times*.

This is a nasty rapid with an 8-foot fall, and I considered myself in more danger than at any point on the river trip. The channel is shaped like an "S" and the swift current drives the boat against a large mass of bed rock about half way down the rapid [now known as "The Island" or "The Bedrock"]. All were frightened for the moment and felt lucky to get through with no mishap. We found the camp at the head of #32 rapids [Dubendorff Rapid], which looked nasty and were glad something had held up the other boats, as it was nearly dark and Tapeats Creek is two miles below. We have to plan our camp sites carefully for one cannot leave the survey line and run on to camp and then walk back the next day. The Supai [15'] map shows the creek at the head of this rapid as a perennial stream, but it was as dry as a bone.[7] However, the next morning we found a fine clear stream 800 feet below, so decided that Evans had the water in the wrong creek. The net result, which is about the only thing we have found to lay up against Evans, was that we had to drink mud from the river when we could have had fine water by an 800-foot carry.

September 8, 1923 — 11,600 ft³/s

BIRDSEYE: Number 32 [Dubendorff Rapid] looked even worse the next morning ... The Kolbs lined the upper half of this rapid in freezing water of January, 1912, [actually December, 1911] and one of Stone's boats capsized with serious injury to the boatman.[8] No previous party had run the rapid successfully[9] and the boys decided it was worse than Soap Creek or Hance. It has a fall of 15 feet in a few hundred yards and is full of rocks and bad holes.

LINT: Looked the rapid over this morning and decided that it would be too dangerous to run it with loaded boats so we portaged the loads of all the boats except the *Boulder*. On account of the way the *Boulder* is built it is very easy to capsize when empty so I decided to take the chance and run the rapid with the loaded boat.

---

7. Two canyons meet the river at Dubendorff Rapid; the upper, Galloway Canyon, is closest to where the expedition camped and is ephemeral. The lower, Stone Creek, is the perennial stream Birdseye was disappointed not to find. We could not verify Birdseye's claim that the problem was with Evans's map.
8. The boatman was Sylvester Dubendorff, who flipped his boat in this rapid and for whom the rapid is named. Unfortunately, the Board on Geographic Names chose to spell his name Deubendorf, which does not honor the man who uttered those immortal words: "I'd like to try that again. I know I can run it!" Julius F. Stone, *Canyon Country: The Romance of a Drop of Water and a Grain of Sand* (New York: G.P. Putnam's Sons, 1932), 95.
9. No one had successfully run Dubendorff Rapid except, of course, George Flavell, whose 1896 run was unknown to the USGS crew.

We had to carry the goods about half a mile, and as four of the men had very weak backs it took us until about 10:30 to finish the job. Emery decided on a channel on the north side of the river but it was badly cut up with boulders. He struck two rocks at the very beginning and one of his oar locks came out of the socket but he made it through without damage to his boat. Freeman thought that he would try it on the south side so he made a run at it and tried to cut through the reverse wave of a boulder and get into a clear channel. He failed and hit everything in the rapid except the channel. Near the foot of the rapid he went down into a big hole below a boulder and the stern of his boat hit a rock. Blake started next with me about 100 feet behind him. We cut through the same reverse wave that Freeman tried to, pulled to the center of the river, and made it through without touching a rock.

BLAKE: [Had Freeman] had one of the old boats or had the boat been loaded, a serious wreck would have occurred. As it was only a small hole was made in the corner of the boat, the two inch oak reinforcing and copper plate protecting the boat from serious injury. We patched up the *Grand* then ate lunch before loading up.

BIRDSEYE, REPRISE: [After Dubendorff Rapid] ran two small riffles (32-a and b) with falls of 2 and 3 feet at miles 70.3 and 71.1. At the lower one the crew of the *Grand* scared up a big horn sheep and La Rue thinks he caught him by movie. Reached the mouth of Tapeats Creek at 3 P.M. and found an almost ice cold stream of clear water running about 60 sec. feet. Stabler wanted to spend the next day scouting Aztec ruins[10] up the creek, La Rue wanted to measure the flow of the creek and the cook wanted to boil up some ham butts, so decided to make a two-night camp, although we were a day behind our schedule. Only after I had made the decision did I find out that the next day was Sunday, so we had a fourth reason for the stop.[11] We set up the radio for the first time since August 30 and heard KHJ fine and clear but with some little interference or "fading out." Heard for the first time that serious trouble was imminent between England and Italy—that some serious disaster had occurred

---

10. The cultural remains were not Aztec, but Ancestral Puebloan (formerly called Anasazi); Helen C. Fairley, *Changing River: Time, Culture, and the Transformation of Landscape in the Grand Canyon* (Tucson: University of Arizona Press, Statistical Research, Inc., Technical Series 79, 2003).
11. The 1923 USGS expedition is certainly not the first to layover at Tapeats Creek (the Stanton expedition of 1890 did so, although they were upstream about a half mile), nor were they the last. For years, it was a popular layover camp. A flood in 2005 rendered it nearly uncampable, and it was closed to all camping in 2006.

in Japan—could not get it all, and, not least in importance, that Washington had won a game [baseball] by a decisive score.[12]

September 9, 1923 — 10,800 ft³/s

LINT: Emery, Blake, Frank Word, Moore, Stabler, and I went up the creek 3 miles to see the source of the much talked of and mysterious "Thunder River."[13] On the way up we found some old cliff dwellers' ruins and I found a fine white arrow head and Blake found part of the jaw bone (with two teeth) of a human. About 2½ miles up from the river the creek forks and we took the left hand branch and followed it up almost half a mile. There we found a large grove of cottonwood trees and all along the creek were beds of maidenhair ferns. This branch of the creek comes out of a hole in the sheer limestone wall and is a regular cataract for it falls 900 feet in about 2,000 feet down to the junction of the two branches. The creek falls 680 feet from the junction down to the river and in places ran three narrow gorges where it was necessary to wade in about 3 feet of swift water. Later on in the day Dodge went up the creek but he took the right-hand branch so missed the waterfalls. We saw tracks and found a small wooden box which showed that someone had been down on the creek some time during the last six months.

BLAKE: As the [Tapeats Creek] canyon widened into a narrow valley there were many evidences of ancient tillers of the soil. Pottery and remains of rock houses were very much in evidence, and a stone dike, built apparently to keep the flood waters of the creek away from the tilled soil, was seen. Several granaries were also found, built in the usual manner, high on the cliffs ... At last we came in sight of the head of the stream. We all felt that we had seen one of the most beautiful sights ever created by nature and could not help but remark upon its being less than a mile from, but outside of the Grand Canyon National Park, the right fork of Tapeats Creek being the west boundary of the park. On the return trip to

---

12. The trouble that Birdseye reports between Italy and England is likely related to the Italian-Greco conflict mentioned by Freeman on August 30. "Mussolini is Victor," *Los Angeles Times* September 8, 1923 (accessed March 1, 2006). A massive earthquake hit Tokyo on September 1, 1923, initiating an enormous fire and killing over 100,000. *Encylopaedia Britannica Online*, "Earthquake," http://search.eb.com/eb/article-60443 (accessed June 13, 2006). The Washington Senators beat the New York Yankees, "Yanks Zeroed by Washington," *Los Angeles Times*, September 9, 1923 (accessed March 1, 2006).

13. The hike from the mouth of Tapeats Creek to Thunder River is another popular hike among the river running community, particularly those on extended trips. Some adventurous hikers continue on over the summit and descend to Deer Creek Falls, thereby bypassing Granite Narrows with their hike.

camp we went bathing in the icy water. Dodge, who had been over the cliffs and had not followed the stream, joined us. He, Frank Word, and I went ahead of Emery and Leigh, and were soon wading the cold water above camp, when suddenly a hundred pound cactus sped by. The plant would have been dangerous without the menace of its sharp spines, so we watched closely for others. They came in numbers, bobbing and bounding in the swift water. We saw the joke instantly, and laughed heartily as we pictured the fun Emery and Leigh were having as they stayed behind and rolled the dangerous missiles into the water. No doubt they pictured the three men ahead hugging the canyon wall and staring up stream in terror, afraid to cross the stream. Indeed we kept a sharp lookout and soon heard a shout from two or three hundred feet above. Looking up we saw two white hats and heard the loud and mocking laughter of Leigh and Emery. We laughed last, however, as they got rimmed and had a good deal of difficulty in getting over the top to where they could start down.

FREEMAN: La Rue measured the creek this afternoon and found it to have a flow of 93 second-feet—two and a half times that of Bright Angel at the time of our visit. Moore won a pool for the closest guess of the flow of the stream.

September 10, 1923 — 10,600 ft³/s

BLAKE: I was the first one thru the rapid [Tapeats Rapid] this morning and had no trouble. A short distance below, however, I pulled sharply toward shore to avoid the big water and under estimated the swiftness of the current, so struck the right corner of the boat on a big rock near shore. The impact knocked Dodge off the stern of the boat, the only damage done being the wetting of his tobacco. The river was swift for a time, and in places was very dangerous, requiring hard pulling to keep from being thrown against the wall where the whole current piled up, making the water at the point of contact several feet higher than it otherwise would have been. There were also long stretches of smooth water where beaches and landing places were few.

BIRDSEYE: At mile 73.1 the granite dips up for about ¾ mile only. At the entrance to this short granite gorge the walls are only about 75 feet apart. We named this point Granite Narrows. Reached the mouth of Deer Creek coming in from the right at mile 74.3. Deer Creek comes out of a cleft in the rock wall in a beautiful falls about 75 feet high. La Rue measured the flow at 8.21 second feet.

The water is clear and cold. About 100 yards below on the right bank a series of springs has quite a bit of flow and is surrounded by green verdure [lots of poison ivy here]. Ran riffle 34-c with a fall of 2 feet at mile 76.1 [Doris Rapid][14] This riffle has a nasty current which drives the boats against the right wall and has boulders near the surface on the left side. All rode through safely but the *Boulder* grazed two rocks in trying to keep to the left out of the current. Ran riffle 34-d with a fall of 3 feet at mile 76.5. Reached Fishtail Canyon at noon. The stream in this canyon—which comes in from the north at mile 77.1—was dry. The eclipse of the sun began at 1 PM, became 75 per cent total at 2:10 PM and was over at about 3:15.[15] It was at no time too dark to continue on survey line. The shadows were fussy and the whole canyon took on a desolate color or aspect. We had several pairs of dark glasses so all got a good view of the eclipse. Freeman took a photo of the sun during the eclipse.

FREEMAN: The eclipse of the sun had set in before we finished lunch, and the work of the afternoon was started in a ghostly fading light ... The weird greenish light which prevailed during most of the eighty percent eclipse was much like that which precedes a cyclone or typhoon. The sensation of running riffles in it was quite uncanny.

BIRDSEYE, REPRISE: Ran Fishtail Canyon Rapids (#35) just below the mouth of the canyon. This rapid has a fall of 10 feet and has some huge waves—one of which nearly knocked me off the deck of the *Boulder*. All boats went through without trouble. Ran a 3-foot riffle (35-a) at mile 77.5. Reached Neighing Horse Canyon at mile 77.8—a dry stream coming in from the south [140 Mile Canyon, also known as Keyhole Canyon]. Ran riffle 35-b with a 5-foot fall

---

14. Doris Rapid, at mile 137.5, is named for Doris Nevills, the wife of pioneering commercial river runner Norman Nevills. In 1940, on Norm's second and Doris's first trip through Grand Canyon, they encountered a rapid much-changed from its condition in 1938. Norm ran without scouting, and Doris was thrown into the water. Comparing the 1923 survey data with 2000 Lidar data, the rapid increased its drop from one foot in 1923 to five–six feet in 2000 as a result of a debris flow that likely occurred between 1938 and 1940.
15. The 1923 expedition was the second trip to have run the canyon during a total solar eclipse that affected the Grand Canyon sun to a large extent, the first being the first Powell expedition, which glimpsed one on August 7, 1869. In neither case did the path of totality cross the skies above Grand Canyon, but Bradley reported that the sun was about half covered before clouds obscured the view, while Birdseye reported the sun about 75 percent covered. The next time a total eclipse will have such an effect will be on August 12, 2045; there will also be an annular solar eclipse on May 20, 2012. NASA, "NASA Eclipse Home Page," http://sunearth.gsfc.nasa.gov/eclipse/eclipse.html (accessed August 31, 2006).

at mile 77.9. For the next 3½ miles to Kanab Creek the water is almost as quiet as a mill pond, although there is considerable current. There are no riffles and the boatmen had to pull steadily to make any speed. We found Kanab Creek coming in from the north at mile 81.6, running about 3½ sec. feet of water, slightly discolored and slightly brackish. The mud bank at the mouth of the creek was soft and afforded poor mooring for the boats, so we camped on the left bank of the river about 1,000 feet above the mouth of the creek. Our day's run was 9.5 miles, the longest yet on this trip. Tried the radio but it would not work for the first time on the trip. Could not tell at night what was the trouble.

September 11, 1923 — 10,500 ft³/s

LINT: Burchard, Birdseye, and Dodge worked 3.1 miles up Kanab Creek. Moore, Freeman, La Rue, and Stabler went up the creek to take pictures and kill time. Emery, Blake, and I crossed the river and took a bath and shaved and then brought back a supply of creek water. Frank stayed in camp and tortured himself by shaving. Had a wind storm in the evening.

BIRDSEYE: Traversed up Kanab Creek ... to the 2100 contour crossing, 218 feet above the river. Found easy going but short sights on account of the winding meander of the stream and precipitous walls of the canyon. Did not get back to camp until after 3 P.M. so decided to remain in the same camp for the night. Attempted to repair the radio and found one unsoldered joint in the wiring. Found that the solder and iron had been lost so made the best connection we could without solder. Also found the B battery badly bulged and fear it is about gone, probably due to the extreme heat of the past few days.

STABLER: The water is better quality and somewhat larger flow (probably 50 per cent greater) a few hundred yards above the mouth [of Kanab Creek]. Some beautiful little pools along the creek and took two cool baths there during the day. Freeman and I, ahead of the rest, saw a duck apparently wounded in the wing. It hid under the rocks and I pulled it out. It was thin and evidently had had hard pickings. The two most interesting sites were a dripping spring and an overhanging wall. The spring covered an area about 150 feet high and 300 feet long. The dripping water left a $CaCO_3$ deposit and the whole area was covered with flowers and ferns. Bees and butterflies were interviewing the flowers. The overhanging rock formed a great cave through which the creek runs, backed by great rock steps and ledges. The overhang is about 100 feet and the height of the overhanging edge about the same. A little shower broke up a bridge game about 8.30 P.M.

September 12, 1923 — 9,730 ft³/s

BLAKE: We ran the Kanab Creek Rapid without trouble. It had 18 feet fall, but was long and had a good channel. Very little slow water was encountered today. Lunch was eaten in the shade of an overhanging cliff of rock. We made five miles before noon and expected to make at least five more in the afternoon, but came only a mile and a quarter when we had to stop and inspect a rough looking rapid [Upset Rapid]. The inspection proved that we had been wise in not running it without looking it over.

LINT: After looking the rapid over we decided to portage the equipment around to the lower end of it so we emptied all the boats with the exception of the *Boulder* which I prefer to run loaded. Made the portage and Emery and I started through. Almost at the foot of the rapid and on the north side is a huge boulder which is just barely submerged and the water falls almost sheer over the lower side of it. This causes the water to roll back in toward the boulder and we knew that to go over the boulder or near it would mean a certain upset and a good chance of the boatmen getting in the "backturn" and drowning. It is an impossibility to run it on the south side on account of three large boulders. Between the large boulder and the north shore is a clear channel about 12 feet wide and this was what we were aiming for.[16] Emery started in the center and tried to cut across to the north side but he didn't quite make it and went over the edge of the boulder and upset. Luckily the boat was carried out of the hole and on down stream. As the boat went over Emery grabbed an oar but that pulled loose and he kept his hand on the rear hatch cover until a wave knocked him back into the cockpit. He was under the boat and under water for about 100 yards ... before he could get out and crawl on top of the upturned boat. I followed Emery about 250 feet but kept to the north shore and dodged the rocks close in. I was about half way through when I saw the *Marble* was upturned so I made the best time that I could, hit the narrow channel OK, and then turned my boat bow downstream and made all the speed that I could. When I was within

---

16. At this water level, most Grand Canyon boaters consider there to be two safe runs in Upset Rapid: the first is to hug the right bank, passing the right side of the hole, and the other is to go left through some of the wildest waves regularly run in Grand Canyon. The 1923 USGS expedition boatmen did not consider the second an option.

about 150 feet of Emery, Dodge swam out and grabbed his painter. I pulled into Dodge and he tied the painter onto the stern life line of my boat and then climbed aboard. I pulled into shore and Dodge snubbed my boat to a rock and I went to see if Emery was injured. He had gotten a little water in his lungs and strangled but he soon got over that. We righted his boat and bailed it out and then lined them both about 100 yards upstream. We then took the boats apart and lined the *Boulder* up to camp and then went back and lined the *Marble* up. The only damage done was the loss of one oar and a pail of lard which was smeared all over inside of the rear hold.

BIRDSEYE: Blake then came through in the *Glen*, and had a hard pull to avoid the big hole by inches, also a hard pull to get ashore at camp. Freeman came last in the *Grand* and made the best run of all except Lint, who always runs best. He kept well to the right at the brink of the falls and pulled so hard that he had ample time to rest on his oars a few seconds before reaching the big hole, passing it by a safe margin and pulling in at camp easily. All were tired and wet and supper was late so there was not much delay in rolling in. Made no attempt to use the radio.

DODGE:[17] Not having a boat to run through a rapid and not wishing to be a bystander if anything could be done to help, and knowing the first boat through was always in danger of cracking up, perhaps injuring the boatman, too, and without aid available, I made a practice of going to the foot of a rapid with a coil of throwing rope. I'd pick a spot I wasn't afraid of diving into if necessary, or a good rope-throwing spot, and wait for the first boat. Once La Rue wandered down to where I was standing and asked, "Why do you always come down here, Frank, when the excitement is up at the top?" Well, I told him it was the only thing I could think of to give the first boat a hand if needed. "What do you mean," he asked. "Would you go after a boat from here to help?" And when I said, "Sure, I would," he turned about without another word and left me. I knew right then that in his mind I was nothing but a whopping liar.[18] Several weeks later I vindicated my assertion. The lead boat went over a huge rock [in Upset Rapid], falling into the hole below, and turned over. A third of a mile below it, there was I looking directly upstream and seeing all that went on. The boatman [Kolb] seemed to be muddled and was doing nothing except floating towards Yuma and points further south. The boat would pass me

---

17. Dodge, *Saga*, 37.
18. See La Rue's entry of August 2.

about seventy feet off, so judging its speed and my swimming, we met where I'd expected to meet. I grabbed for the long painter and started back to shore. And was I a hero! That night Birdseye came to me and said, "I think you're worth a boatman's wages after this. From now on you're getting $200.00." Don't let the above make you think I thought I was a hero. Hell no; but I've never got over the attitude of non-swimmers to what is possible and what isn't. The least little agitation to the water scares 'em—as it should—but they keep thinking it's dangerous for all.

FREEMAN: In consideration of the rough work of running, portaging and lining, the Chief drew on the medicine chest for the snake-bite medicine and prescribed a liberal dose to all hands.[19]

September 13, 1923 — 11,800 ft³/s

BLAKE: We were up early this morning. After breakfast the loading of the boats commenced. Emery took a movie of the place where he upset. Freeman and La Rue starred in one of their little cave-man scenes, the beginning of which took place at Lee's Ferry, when they persuaded Edith Kolb to dress as a cave woman and take part in a rough and tumble scene with Freeman.

FREEMAN: Two mountain sheep—ewes—appeared high up on the cliff of the opposite side during breakfast.[20] They were very curious, and rather nervous than frightened. The water was swift for half a mile below the rapid, and then alternated with quiet reaches and light riffles all the way to Havasu. With walls almost sheer to the water from the top of the Supai, the canyon presented a finer scenic spectacle from the river than in any section through which we have passed below the Little Colorado. We continued on to the mouth of Havasu Creek without stopping for lunch. Havasu Creek is the largest flowing into the Grand Canyon. The creek itself takes its rise in springs in the Havasupai Indian Reservation below the outer rim, probably flowing from under the Redwall. There are two or three fine falls near the Indian village, and many beautiful pools. Normally the water is of a brilliant blue in reflected light, but at present is of a cinnabar color from the stains of earth washed down in the recent rains.

BIRDSEYE: Reached the mouth of Havasu Creek at mile 95.0 and found the stream running red from yesterday's storm. La Rue

---

19. The "snake-bite medicine" likely was a euphemism for a flask of Prohibition-banned liquor.
20. Bighorn sheep remain a common sight in this part of Grand Canyon.

measured the flow as 75 sec. feet. Checked on Evans's V.A.B.M. [Vertical Angle Bench Mark] with an error of +4.3 feet, finding the water elevation 1782, or 14 feet lower than Evans reported on July 10, 1923. Roger Birdseye and Charles Fisk were waiting with supplies to last until we reached Diamond Creek. They brought in about 700 lbs., using 19 Supai Indians as packers for the last 7 miles down over the falls and narrow gorge of Havasu Creek. The Indians would not pack over 40 lbs. and required a good deal of coaxing to come at all. None of them had ever been to the river before—nor seen a boat.

LA RUE:[21] One Indian 90 years old but he packed his 40 lbs. over the precipitous walls and kept up with the younger men of his tribe.

BIRDSEYE, REPRISE: Frank Word left at this point, having suffered with the glare of the sun on his eyes so he could hardly see. He had told us at Bass that he would have to go out. Roger brought in Felix Koms [Kominsky], a 200-lb. baby who seems to be both a good cook and jolly fellow. He is wearing a straw sailor hat which the boys vow to lose in the first rapid.[22]

BLAKE, AUTOBIOGRAPHY:[23] On the way down he [Felix] had sat in a prickly pear clump of cactus, and was still picking out spines at every opportunity. The Indians had hugely enjoyed the spectacle of the heavy Pole taking downs his pants and looking for stickers, whenever they stopped to rest.

KOLB, LETTER:[24] Supai Sep 13, Well dear, It was nice to get here to the mouth of Supai and familiar faces ... Well this is a fat cook I have to carry from here on. I would like to have been here when the 19 Indians carried the stuff in, it would have made a great picture ... No I don't want you to come in at Diamond Creek. It is too hard a trip and as for later you better make no plans until we see later. I might have to come home to work up some photos and might take the train home. Give Frank the cook one of our books and if there is a couple pictures you want to give him it is ok with me. We all feel sorry to see him leave. I had an upset last evening

---

21. La Rue, photographic diary, 38–39.
22. Felix Kominsky is as mysterious a man as the one he replaced as cook. He was apparently uncertain of the spelling of his own name, which might have been Kominsky, Kaminski, or another variation. According to one source, he pronounced the shortened version of his name "Kooms." His hometown was likely one of the Polish strongholds within Pennsylvania. How he landed in Flagstaff is not known. Undated handwritten note, box 280, folder 35, Marston Collection.
23. Blake, "As I Remember," 169.
24. Emery Kolb to Blanche Kolb, September 13, 1923, box 5, folder 623, Kolb Collection.

but all clear sailing below the back lash. I hope Edith gets strong and don't think it would be a bad idea for you to stay in Flag with her a little ... With love, Emery.

FREEMAN, REPRISE: A heavy thunderstorm just before dark was followed by a rapid rising of the river, which still continues

WORD, LETTER:[25] ... theire is something i am going to tell you you know theire was quite a Bit of grumbling amongst the bunch while we were on the trip i wanted to tell you before i left but never quite had the chance that is one reason why i left. Emery Kolb was the cause of all the trouble he thought you and Mr Freeman were making those movie Pictures for comercial use of course i did suffer with my eyes and do yet but theire continual grumbling and complaning all the time got me dissatisfied he even went so far as to try and get Mr. Birdseye to send you and Mr. Freeman out at Hans trail at least he told me and the Boys so i am satisfied that what he does dosent make any difrence to you but i Just thought you would like to know well i am glad you got back safe and hoping this finds you enjoying good health i am verry resp, Frank Word. P.S. If i can have some of thos pictures i would surely apreciate them would like to come up and have a talk with you some time if i can find time.

September 14, 1923 — 11,700 ft³/s

BIRDSEYE: Roger, Chas. Fisk and Shirley left at 9 A.M., packing out maps, notes, film and radio outfit. The radio set was out of commission due to run-down batteries and one unsoldered connection. Thought best to lighten our boat loads and also to have radio set overhauled and brought in at Diamond Creek.[26] Found we had excess of flour and sugar so abandoned 50 lbs. of each and also one of the large mess boxes which had been broken so was not water-tight. Remained in camp all day, repairing boats, rearranging loads, etc. Burchard made a detailed survey of dam site #14 just above the mouth of Havasu Creek and found one of the best sections on the river for a low power dam—155 feet long at the water and 360 feet across the top at a point 235 feet above water.

BLAKE: Leigh spent most of the day repairing his boat so it would hold up until the end of the trip. I mended shoes and repaired my

---

25. After the trip, Frank Word wrote a letter to Eugene C. La Rue; Frank Word to E. C. La Rue, November 2, 1923, box 2, folder 1, La Rue Collection. In his unique and barely literate way, Word summed up some of the interpersonal difficulties of the trip.
26. The decision to have the radio sent out for repair, instead of attempting a field fix, would have consequences later on as the outside world tried to transmit word of the impending flood.

boat, and helped Leigh with his boat. I had a headache and slept for an hour or two before supper. Our new cook is a wonder. He made a cake in a Dutch oven, and never got out of humor when the wind blew sand and ashes all over the place.

FREEMAN: The river continued to rise during the night, and is now up over two feet this morning. We had to get up at daylight and bring the boats in, as the rock to which they were tied had become submerged.

September 15, 1923 — 10,900 ft³/s

FREEMAN: The river has fallen another foot during the night and is now just about at the stage at which we landed. The river rose slightly for an hour before dark.[27]

LINT: Had to line the boats upstream about 100 yards so as to be able to pull out into the center of the river and avoid some boulders at the mouth of Havasupai [Havasu] Creek. There is a small rapid (No. 40) with 4 feet fall here but it has some rocks in it on the south side. There is a clear channel in the center and on the north side. Made four miles by noon. Rapid No. 41 has 4 feet fall and is 7 miles below Supai Creek [164 Mile Rapid at Tuckup Canyon]. It is a narrow rapid with three rocks in it and the current runs into the wall on the south side of the river. Entered the V on the north side and kept pulling that way. No. 41 is at the mouth of a creek and we worked up it about half a mile. 8.2 miles today and camped on the north side of the river 1.2 miles below Rapid No. 41. I pulled into camp at 6.30 just as the cook yelled "Come and get 'er." Least fall of the river today that we have had in any one day's run since leaving Lee's Ferry's very few whirls.

STABLER: We got an early start today and made 4 miles by noon. Comparatively still water with minor riffles are the rule with walls of red limestone 1200 to 1500 feet almost straight up from the river's edge, some places overhanging. This portion of the canyon is wonderfully picturesque and beautiful. Far more so than the granite gorge above. We now have no good maps and it is difficult to locate ourselves by the old Powell surveys which are on a small scale. Our camp tonight is on a large sand bar on the left bank of the river about a third of a mile below a creek that I think I have properly located on the Powell map—about 7 miles below our starting point

---

27. Unlike the other diarists, Freeman always started his daily entry with a note about what the river was doing. His observations show how little warning the expedition was to have about the impending flood, which would catch them a few miles downstream.

for the day. I have been practicing the semaphore alphabet with Emery and can now converse slowly with it. It is a most helpful thing on a trip like this and permits conversation between persons nearly half a mile apart.[28] Felix, the new cook, is taking hold well. He has not yet been over any real rapids but takes the smaller ones with a smile—as in fact he takes pretty much everything. After supper set a pile of driftwood on fire. It illumined the canyon for two or three hours.

September 16, 1923 — 10,800 ft^3/s

BIRDSEYE: At mile 104.7 came to a deep canyon from the left [National Canyon]. This is shown as Cataract Creek on the Kaibab topographic map and it is my belief that Powell originally applied the name to this creek and not to the creek in Havasu Canyon, although the Kaibab map has been assumed to be in error and for many years Havasu Creek has been known locally as Cataract Creek. The Kaibab map is hard to follow in comparison to the recent maps by Evans and Matthes, but I am sure of the location of Cataract Creek. At mile 106.3 came to a deep canyon from the right [Fern Glen Canyon]—dry at the mouth but several springs with fair drinking water ¼ mile up. Traversed up to elevation 1796 and found the canyon blocked by a boulder slide. Ran #42 rapid at mile 106.3. This is a small rapid with a fall of only 6 feet but has a bad rock and hole in the center at the brink [Fern Glen Rapid]. Have been suffering from ulcers on my back which started from infection in a small boil which developed at Hance Trail on August 16. Three ulcers developed and necrosis of the tissue developed. Had fresh dressings of zinc ointment twice a day from Aug. 23 to September 16. At this point the third ulcer had healed sufficiently to permit of dry dressing and expect no further trouble. The medicine chest has been of particular use to me and Emery Kolb has been a willing and efficient M.D.

BLAKE: About a mile below camp we came to a side canyon on left which had to be surveyed [National Canyon]. Emery and I went up the canyon, as did everyone else. There is a nice stream of water in the upper part, which sinks away when within five or six hundred feet of the river. About a half mile up the canyon narrows into a gorge with a deep pool at the lower end. Emery and I climbed around the pool and by helping each other made our way

---

28. Kolb, in his conflict with Birdseye, would include his knowledge of semaphore as another reason he was owed greater consideration.

up over some boulders which had lodged in the narrow gorge. We then had good going for a thousand feet, when we came to a huge rock which again arrested our progress and which made a waterfall about twenty feet high. After taking a picture of this we climbed around it and explored the canyon for a couple thousand feet more, then returned, taking a bath in the clear pool which stopped the others from going farther. We then sent down to the river and continued work for a mile and a quarter, where we ate lunch at the mouth of a side canyon [Fern Glen Canyon] on the right.

LINT: Blake hit a rock as he was landing after lunch and it drove the extra oars back and split his forward coaming all to pieces. Made 5.2 miles today and camped on the south side of the river.

September 17, 1923 — 9,380 ft³/s

LINT: About three fourths of a mile below camp and on the north side of the river the Colonel and Burchard worked 1½ miles up a side canyon [Stairway Canyon]. A quarter of a mile below this canyon and on the south side of the river is another long canyon [Mohawk Canyon] and they worked 1½ miles up it. Rapid No. 43 is at the mouth of this canyon. It is a long rapid with 10 feet fall but is nothing but a straight chute with large waves. At the head of it, in the center of the river, is one large boulder. Ate lunch at the head of the rapid. Felix got a pail of water and started to step from the stern of the *Glen* onto the stern of the *Marble*. The boats drifted apart and there he was with one foot on one boat and the other foot on the other boat. The boats kept drifting apart and consequently Felix went into the river and the war whoops that he let out would have done credit to an Indian. 3.9 miles today and camped on the north side of the river at the mouth of a canyon [Cove Canyon]. Burchard and the Colonel worked about half a mile up this canyon. Rained this afternoon.

STABLER: Both creeks [Stairway and Mohawk Canyons] were dry at the river but the one on the north had pools of fair water and the one on the south a small running stream perhaps a third of a mile back from the river. We lunched at the creek on the south. Below this stream was a rather sharp rapid [Gateway Rapid] in which Felix lost his straw hat. It floated and we recovered it below.

BLAKE: Emery caught a bonytail fish today, the first caught on the trip.[29] From camp we can see a tent on a high talus slope. It is

---

29. This likely was a humpback chub, an endangered species that warrants much concern now.

supposed to be the tent of the silver mine which we expected to find farther down the river.[30] We set a big drift of wood on fire so if there was anyone at the mine they would know of our presence.

September 18, 1923 — 42,800 ft³/s

[The crew had experienced rain, or at least noticed passing thunderstorms, for the previous several days. Without their radio, the men could not know that a massive flood was headed their way.]

FREEMAN: The river is down a few inches. A half mile above the lava we sighted a large mountain ram, watching us from a rock half way up the cliff. Undisturbed by our shouts, he remained motionless until our boats passed out of sight.

BIRDSEYE: At mile 116.0 we passed a lava pinnacle in the center of the river channel [Vulcan's Anvil]. It is cylindrical in shape, about 50 feet in diameter and 60 feet high. Dr. Moore says this is probably not a volcanic plug but that the rest of the barrier has probably been worn away. We named this pinnacle "Lava Rock Island." (This is an unmistakable landmark and boat travelers should take warning of an "un-runable" rapids (Lava Falls) 1½ miles below.) At mile 117.6 we reached a canyon from the left caused by Toroweap fault [Prospect Canyon]. This canyon is dry and the ascent steep. The beds on the west side of the canyon show a downward displacement of about 580 feet. Erosion has cut deeply into the alluvial fan at the mouth of the canyon. Traversed up to elevation 2260 feet. At the mouth of this canyon (at mile 117.6) we came to Lava Falls which we consider the worst rapid so far in the Marble and Grand Canyon series. It has a fall of only 10 feet in medium low water stage[31] but this fall is very steep and rocky, with no channel through which boats can be run. In high water one might run through but small boats would surely upset in the huge waves which in high water stage run for ¼ mile below. No one has ever run this rapid successfully.[32] Warm springs with fair drinking water enter the river all along the left bank from the brink of the falls to a point ¼ mile below. There are three fair-sized streams and a number of smaller ones, the estimated flow of all being 15 sec. feet. A large area between the river and cliff is covered by tule grass and thistles, in which there is a large

---

30. The tent probably was at the Little Chicken mine; Billingsley, Spamer, and Menkes, *Quest for the Pillar of Gold*, 40.
31. Birdseye mentions the drop through Lava Falls Rapid in 1923 as ten feet before the flood. See note 35.
32. Birdseye was unaware of George Flavell's successful run in 1896, when he claimed the lively rapid put eight inches of water in his boat; Flavell, *Log of the Panthon*, 69.

pool of clear water. The water is evidently heavily impregnated with lime or other mineral (possibly soda) as small terraces have been built up in places clear to the water's edge.

STABLER: The canyon here is right on the Toroweap fault, Tapeats Sandstone coming to the surface of the water just above while Muav Limestone is at the water just below. The north or right wall is a sheer cliff composed of a series of lava flows while the left bank is a great boulder and gravel fill a hundred feet in height running back to the limestones. The cinder cone on the right bank mentioned by others is not in evidence though the lava flows are eroded into a somewhat conical butte just at the foot of the falls.

BLAKE: We had a long stretch of fast water today, and only one small rapid, until noon, when we ate lunch at the head of Lava Falls, which has a steep fall for about two hundred feet, and is strewn with boulders, which on the south side for about two-thirds of the way across the river, protrude above the water. The rapid is evidently made by a lava dike which has flowed down the cliffs from the north, and probably dammed the river at one time to a much greater height.[33]

LINT: Lava Falls has a fall of 10 feet. Across the head of the rapid is a row of rocks in the shape of a quarter circle. Then below this row are numerous other rocks all through the rapid. Rapid No. 44. At the stage of water that we had today there was no possible way of getting past the first line of rocks. If a person could get through them there is a narrow but clear channel on down the center of the rapid.

FREEMAN, REPRISE: It was decided to portage all the loads and to line the boats down the left bank, lifting them over boulders where it seemed too rough to let them go outside. This was successfully accomplished, Lint riding all four boats and fending off very effectively. The boats are all down and reloaded, and we are camped at the head of the rapid in the sand and boulders. Birdseye and Felix fell into the water during lining operations, the former while La Rue was running the movie. If it was prearranged, the Colonel is to be congratulated on the fervent registration of passing emotions.

---

33. Blake repeats an often-quoted origin for Lava Falls—it is the long-eroded remnant of a lava dam. We now know that the rapid is formed from frequent debris flows from Prospect Canyon, which Birdseye notes reaches the Colorado River at the top of the rapid. This canyon has produced the most debris flows of any within Grand Canyon during the twentieth century; R. H. Webb, T. S. Melis, P. G. Griffiths, J. G. Elliott, T. E. Cerling, R. J. Poreda, T. W. Wise, and J. E. Pizzuto, *Lava Falls Rapid in Grand Canyon: Effects of Late Holocene Debris Flows on the Colorado River* (U.S. Geological Survey Professional Paper 1591, 1999).

At one time the whole river must have been dammed with the lava flow, but this obstruction, extending for many miles, has gradually been cut down. Lava Falls is not the remains of this dam, but is plainly the result of later slides from the right wall, which extended at one time all the way across the river and against the delta built out by a large unnamed creek from the left. There are abrupt drops of from four to six feet in places, and those are doubtless increased as the river approaches it slower stages. The rapid is far from being a fall in the generally accepted sense of the term, however, any more than those of Cataract Canyon are cataracts. Powell's nomenclature of rapids was very often calculated for effect.

BIRDSEYE, REPRISE: At 6:30 P.M. the river commenced to rise and continued to rise all night at the rate of about 1½ foot per hour. At 8 PM the boats were pounding so badly that quick action became imperative. Kolb thought he remembered a beach ¼ mile below, so he and Lint pushed off in the *Boulder* to reconnoitre. It was a risky undertaking as Kolb was not sure of the beach and the river was pounding the rocks along the shore.

LINT, REPRISE: We had a carbide lamp and a flashlight but it was hard going as it was dark. We found a small beach about 2000 feet below camp so we tied my boat up and went back up to the others. There are many warm springs along the bank and the vegetation is thick and we had quite a time making it back. We then pulled Emery's boat up on the rocks about 10 feet above the water. Emery, Blake, Freeman, and I then took the *Glen* and the *Grand* down to where my boat was. Freeman then went back to camp (I guess he didn't like our company).

FREEMAN, ENTRY FROM SEPTEMBER 19: Beaching the *Grand* and the *Glen* besides the *Boulder*, I set off up the bank with a lantern to return to camp. The others had brought their beds, but there had been no chance to get mine in the rush of pushing off. It was vile work pushing across the slushy, crumbling springs formation, with a tangle of rotting vegetation underfoot, and huge thistle and saw-edged grasses growing higher than my head. These pricklers and cutters were particularly unpleasant as a consequence of the fact that my back and legs were bare. Failing to make any headway back along the cliff, I returned to the boats and started again, this time closer to the bank. After fording three or four warm streams, I finally came to opener going, but here were a half dozen breeds of cacti to complicate navigation. The last hundred yards was along the cliff and over the boulders, among which the rising flood was beginning to swirl. The blind ride in the dark down the side of the rapid was preferable to the return by land a dozen times over.

BIRDSEYE, REPRISE: The boys below did not sleep any as they were busy all night pulling up first one boat and then another. At 11 P.M. the *Marble* was afloat and all hands were called to pull her up an additional 8 feet. It was difficult and dangerous work with the waves lashing the rocks so as to prevent footing below. Felix went to bed 10 feet above the water but was flooded out at midnight and we had to rescue his bed and clothes. During the night we moved the beds and cook outfit 3 times. None of us got much sleep.

BLAKE, REPRISE: We had built a fire in order to have light, and as it was nearing midnight we felt the need of refreshments, so I found the proper ingredients and made a gallon or more of cocoa. After pulling the boats up again we went to bed, but did not dare to go to sleep as the river was still rising, and the position of the boats had to be changed constantly. As it was they were being badly battered. About 3 o'clock we decided it was safe to go to bed, and after piling heavy rocks upon the prow of each boat to keep them steady, we turned in. We had to get up only once again before daylight to attend to the boats.

FREEMAN, FROM *DOWN THE GRAND CANYON*:[34] If the radio set up had been with us we would have received numerous messages broadcast from several stations in response to wires from Washington advising us that one of the heaviest storms of recent years had broken upon the basin of the Little Colorado, and warning us to be on the lookout for the waters of a very heavy flood. Having received such warnings, we should unquestionably have picked out a broad open section, with ample room to back away from a rise, and waited for the flood to pass. Unwarned, we were surprised at a time and place that were far from favourable—twilight on the brink of Lava Falls.

September 19, 1923 — 98,500 ft³/s

BIRDSEYE, ENTRY FROM SEPTEMBER 18: I turned in at 2:30 A.M. and got up at 4:30 but slept little. Had placed my bed on a flat rock 20 feet above low water and at 4:30 the spray was lashing my rock. At daylight I saw that the *Marble* was safe for an additional rise of 2 feet and went below to find the other 3 boats safe but beginning to pound in the waves. Called all hands from the camp above and we pulled the 3 boats up high and dry. The 3 boys who had stayed below were stiff and sore with hands lacerated from pulling on the ropes all night.

---

34. Freeman, *Down the Grand Canyon*, 358.

Downstream view of the *Marble*, winched up the bank at Lava Falls Rapid during the September 18 flood in the Colorado River.

E. C. La Rue 316, courtesy of the U.S. Geological Survey Photograph Library.

On September 22, after the floodwaters had receded, E. C. La Rue photographed the *Marble*, stranded high above water where it had been winched up several days before. In this photo, from left to right, are Birdseye, Lint, Blake, and Kolb.

E. C. La Rue 318, courtesy of the U.S. Geological Survey Photograph Library.

BLAKE: When we awoke at sunup we found that someone had pulled the boats up while we slept, and soon learned from Dodge who came down after supplies, for breakfast, that the colonel had been down about an hour before.

BIRDSEYE: At 8 AM the river had risen 16 feet and continued to rise slowly all day, reaching the peak of a 20-foot rise at about 6 PM.[35] We portaged the rest of the equipment to the cove below and prepared for camping several days. It being impossible to launch the *Marble*, we pulled her up again several feet at 10 A.M. and spent the rest of the day washing selves and clothing in the warm springs. Both nights were chilly—60° Fahr.—so the warm baths in water about 90° were pleasant. The rapids had an entirely different look in high water. Immense quantities of drift with many large logs floated by during the day and some were tossed completely out of the water by the large waves. We estimated the rise measured in volume to be about 60,000 sec. feet.[36] Evidently the frequent rains from September 13 to 16 were accompanied by worse conditions up-river, but whether or not the flood came from the Little Colorado River or from still further up stream we are unable to tell. Kept close watch on the river until it was apparent that the peak rise had been reached, then everyone turned in for a good sleep.

FREEMAN: The Colorado sprung one of its notorious surprises on us last night, and is still trying to keep up its joke this morning. A distinctly unpleasant odor to the water suggested that the flood was coming from the Little Colorado, the founts of which we had already discovered were olfactorily offensive. The rapid had altered in character overnight beyond belief. The head of the dam of boulders, where yesterday there had been a broken series of abrupt falls, was now completely submerged—drowned out. Over it rushed a broad, solid chute of wildly running water which did not begin to

---

35. This abrupt rise led to one of the myths about Lava Falls that has been attributed to the 1923 USGS expedition. Some printed river guides portray Lava Falls as having a 37-foot drop; Larry Stevens, *Grand Canyon: A Guidebook* (Flagstaff, AZ: Red Lake Books, 1983). This is in fact the drop from the top of the rapid to the bottom of the fifth secondary rapid one and one-half miles downstream. In actuality, and depending on river flow, the drop through the initial drop of Lava Falls is only about fourteen feet; Webb and others, *Lava Falls*. As mentioned previously, the 1923 USGS expedition found a drop of ten feet here before the flood; the difference is explained by multiple debris flows from 1939 through 1995.

36. The peak discharge for this flood was 120,000 ft^3/s on the Little Colorado River and 115,000 ft^3/s on the Colorado River at Phantom Ranch. The flood peak decreased downstream from its source in the Little Colorado River basin and no doubt was less at Lava Falls, so we have no way of estimating what the peak discharge might have been here, but it likely was much larger than Birdseye's estimate.

break into waves until half way down what had formerly been the rapid. From there on the waves were tremendous—quite the largest we have seen. These culminated in a great comber just above where our boats are pulled up—a point where yesterday there was only hard-running but comparatively smooth water. This wave is an enormous boil or fountain, which at times must measure from 15 to 20 feet from trough to crest. It appears to be caused by conflicting currents rather than by a rock, and rarely assumes the same form twice in succession. At times, in breaking back, it cups down a large quantity of air, which, when compressed, throws out a jet of spume like that from a cavernous blow-hole on the ocean shore.

LINT: Along in the afternoon we took the blocks and tackle up and pulled Emery's boat up about 6 feet. The bank is almost straight up and down so we had to use something besides straight muscle. La Rue found a rattlesnake and, being afraid of it, called on Emery for help. Emery got the snake out where La Rue could take some movies[37] and then he killed it. After it was all over with La Rue said "This is worse than running a big rapid, I'm shaking like a leaf." Which goes to show that he is afraid of water.

STABLER: The character of the falls has completely changed. Instead of a short sharp cataract we now have a rapid a mile long with huge waves 15 to 20 feet high. It looks now as though the chances would be considerably in favor of a boat running it successfully. However, we are well located and will stay here until the river subsides to reasonable stage.

September 20, 1923 — 87,800 ft³/s

BIRDSEYE: The river seemed to have receded a foot or so but there was no apparent change during the day. The *Marble* was in a safe position but it was still impossible to launch her. Our instrument station was still several feet under water and the river too rough for navigation so we called the day Sunday, having worked all day on the sixteenth.

BLAKE: The river dropped two or three feet during the night. Leigh and I spent the morning repairing the bottom of my boat. Freeman and Emery helped us after noon. The bottom of the boat is so rotten that wooden pegs can be driven easily thru it. Emery and I played Moore and Stabler a game of five hundred which was interrupted by the cook's call for supper.

---

37. This footage was included in the film that the USGS made of the expedition. US Geological Survey, *The Survey of the Grand Canyon of the Colorado*.

September 21, 1923 — 47,800 ft³/s

FREEMAN: The river has fallen six feet during the night, but is still ten or twelve feet above the stage of the day of our arrival. Kolb considers it too rough to launch and load the boats. There is a renewed run of drift, but this is probably drawn out of the eddies by the fall, and is not an indication of another rise.

BLAKE: Emery and I started to walk to the top of the side canyon above the rapid, about nine o'clock. We had a fairly good going over the boulders of the bottom of the wash for a half mile. We filled our small canteen from the trickle of water that flows for a few hundred yards before it disappears in the gravel. Near the head of the canyon the slope rises very swiftly and is almost impassable. We climbed over masses of solidly cemented boulders for a time, then around the jagged face of a cliff to a more promising looking crevice filled with lava boulders. Once or twice a rock would slip or move slightly, causing hundreds of tons of boulders to creep an inch or so, making us wish we were elsewhere but it being fully as dangerous to retreat as to proceed, we kept on toward the skyline. Finally, after several hours of climbing which had been retarded by the taking of pictures of some of the most striking views, we reached a bench about two hundred feet below the rim. Here we thought an easier route might be found to the left and so followed the shelf for a few hundred feet, but decided there was no route in that direction, so retraced our steps and continued our climb to the top. Here we found the lava covered with gravelly soil well sodded to gramma grass which seemed not to have been stocked for some time, the old horse tracks were numerous.[38] We were not on top, however, but were in a wide, gently sloping valley, with sheer cliffs to the east and a more gradually sloping, tho high hill to the west. The valley was approximately two or three miles wide and several times as long. A high cinder cone rose on the east side of the valley about a mile south of the rim where we climbed out. Various side draws came down from the cliffs and we followed up one of these in search of water where we found evidences of an old campsite and a wooden packbox. Many deer tracks were seen also. About a half hour after arriving on top Emery suddenly stopped and listened, and then said, "I hear an automobile." I listened and could hear a dull buzz, but thought it was the wind. Emery insisted that

---

38. Likely these were burro trails, not horse tracks. Burros, which prospectors released in western Grand Canyon beginning in the 1890s, proliferated until they were removed by the National Park Service in the early 1980s.

it was not the wind, and presently we both exclaimed, as the sound became very distinct, "It's an airplane." We then searched the sky, and as the sound became very plain we caught sight of a big plane which we judged to be two thousand feet in the air, headed northeast across the canyon. We thought it might be a scouting plane, searching for our party to see if the sudden rise of the river had damaged us, so we waved our white hats, but could attract no attention. The plane was not, however, following the course of the canyon, so it was decided that it had nothing to do with our party.[39]

As we started back to camp we looked for an easier route for descent than that we had used in our climb, so after viewing all possibilities, we decided on a route a half mile to the west which would take us around onto a cinder slide which reached to within a hundred feet of the bottom of the wash. This route we followed, tho it was pretty dangerous, until we reached the cinder talus which was as steep as the volcanic cinders would lie. Our descent was very similar to roller skating, as each step would take a couple of yards and we had to keep our bodies bent forward so as to keep up with our feet. We descended about a thousand feet in this manner then crossed a lava ridge and down a short cinder slide on the other side, from where we worked our way over the rocky slopes and ledges to the bottom of the wash, from where we had fair going to the river. We bathed in the supposedly warm spring near camp, and arrived in time for a short rest before supper. We had taken no lunch and had only a quart of water between us during the day, so were rather hungry. I played cards until about nine o'clock and then slept like a log all night.

STABLER: The river has been falling all day and should be O.K. for travel tomorrow—in fact, is all right now. Moore and I spent the morning speculating on the lava dam at this point.[40] While browsing about we located what must be the cinder cone mentioned by others. Could not be sure whether it was one. Also found where seven Indian houses had been under a cliff and picked up one piece of pottery and saw a few bones.

---

39. In the National Archives copy of Blake's diary is the following note: "The plane was an Army Air Force Plane on an aerial photography mission flown by Capt. A. W. Stevens of McCook Field, Dayton Ohio.—Nellie C. Carico, September 12, 1966—Special Asst. to Topog. Div. Adm. Officer USGS." Now, tourist aircraft fly a route that takes them over Lava Falls regularly, and the sound of an aircraft would not attract attention.
40. The lava dams at this site, first noted by Powell in 1869 and every river runner since, were studied intensively by Kenneth Hamblin. Kenneth Hamblin, *Late Cenozoic Lava Dams in the Western Grand Canyon* (Boulder, CO: Geological Society of America Memoir 183, 1994).

September 22, 1923 — 26,100 ft³/s

BIRDSEYE: ... resumed our voyage at 9:30 AM with the river still about 7 feet above normal low water stage. Lava Falls was still rolling huge waves which continued for 1/3 mile below the falls. A swift current carried against the left wall below our camp which is so precipitous that one can not walk around. All rode and got a good soaking.

LINT: Launched Emery's boat this morning and he took it down to camp. We cut some rollers and put the other three boats in the water, loaded up, and started our work down the river. Saw several places where the lava flowed down and filled up the small side canyons which were originally cut into the red sandstone. There is more lava on the north side of the river than there is on the south and it is most prominent in the canyons. This morning the water level was 13 feet below what it was at the peak of the flood and it went down about 1.4 feet during the day. No rapids today but numerous riffles, and it was hard to handle the boats and make landings. 7.5 miles today and camped on the south side of the river at the mouth of a small box canyon [possibly about two miles upstream from Whitmore Canyon].

FREEMAN: The increased weight of the water was evident the moment we pushed off, and when it began to slop aboard in running the rapid under the cliff below it proved to be a veritable liquid mud. A splash of it left a white coating on the skin as it dried, while the effect on the eyes was almost blinding. There was the same heaviness to the water we have noticed in previous rises, only much worse, so that the boats were a good deal more unmanageable. The run of water from riffle to riffle was almost continuous, and it was not always easy to pull out into an eddy for a landing. We made about four miles this morning, lunching on a bar above the mouth of a canyon sometimes called Prospect Creek. The fall has been nearly 10 feet to the mile, compensating somewhat for the slow water above Lava Falls.

STABLER: We are camping tonight on a sand and gravel bar on the left bank of the river. Springs in bar—probably seeps from water imprisoned on the last rise—give us excellent clear water. This is gratifying for the river is still full of slimy mud and the water does not clear up readily on settling. Am having a lot of fun discussing geologic and physiographic problems present in the canyon with Moore. He is well informed and I think enjoys greatly working out the problems as they arise.

September 23, 1923 — 17,700 ft³/s

[By this time, the alarm had been sounded nationwide that the 1923 USGS expedition might be in trouble due to the flood. An overturned boat believed to be marked USGS on the bottom was spotted floating near the Katherine Mine, just above present-day Davis Dam. A Kingman reporter posted a September 21 report that made the Associated Press wire, and newspapers across the country picked up the story. The *Los Angeles Times* front-page story on September 23 sported the headline, "WILL SEEK CANYON PARTY, Patrol to Start Search Today for Men Believed to be Battling Swollen Waters Along Colorado." The story included comments from Mabel La Rue, who was "unalarmed and calm;" Walter Mendenhall, USGS chief geologist, who also expressed his doubts about trouble for the party; and Frank Word, ex-cook for the expedition, whose opinion follows.]

WORD, NEWSPAPER INTERVIEW:[41] "Emery Kolb, chief boatman, used to examine all rapids carefully and if one looked dangerous would not make a trial trip through himself," Word declared. "If he was not convinced that the rest of the boats could get through safely, he insisted on a portage being made over the dangerous rapid." The possibility that the wrecked boat might be that of Kolb, was admitted by Word. Such a wreck happened once before, he said, at the Scheynemo Rapids [Word presumably meant Upset Rapid, not Shinumo]. At this time the chief boatman was dragged 300 feet under the boat and was rescued with difficulty. "Unless there was an unusual flood or a cloudburst of greater violence than has been experienced in the district for some time, I do not see how it would be possible for the entire expedition to be wiped out."

[Unaware of the speculation on their fate, the expedition continued its work.]

BIRDSEYE: River fell 1.4 ft. during the night. Still 3.2 feet above normal low water stage. Surveyed geologic section B at camp (scale 1:31,680). Ran rapids #45 [Whitmore Rapid] with a fall of 6 ft. at mile 126.3. Easy to run.

STABLER: Had a rather busy day. After breakfast, I walked a mile along the shore. Then took to the boat and ran a little rapid stopping in the eddy below to inspect a canyon that would have to be surveyed. It ran back—and up—three quarters of a mile and Emery and I explored it to the end. Half way up a considerable fault was

---

41. "Will Seek Canyon Party," *Los Angeles Times*, September 23, 1923 (accessed March 1, 2006).

cut and followed by it—several hundred feet apparently being the Redwall down to Old Snuffy of B.A. Moore followed up with the survey and took notes on it. Half a mile below we camped for lunch and a mile farther we made night camp [unnamed canyon at mile 189.7]. Burchard ran up Prospect Canyon No. 2 while I spent the afternoon recording for Birdseye and a dam site survey. No evidence of Girand's 4-foot monument was visible, but this is evidently his upper dam site. La Rue would have passed without survey though he took a picture. He condemns it from a distance on the basis of supposed shattered rock which I could not find on close inspection. There is, however, a slight slip—2 feet—visible in the Tapeats a short distance above—almost at the point where the survey begins. While by no means a superfine dam site it would serve very well to fill in between Diamond Creek and Havasu, the top of the Tapeats here being about water level at Havasu. Granite—pinkish—comes in here for a mile or more and would form the foundation and lower 80 feet of abutments for the dam. Presumably the granite will disappear at Hurricane fault less than 2 miles below.

FREEMAN: There is a large clear spring at the water's edge just below camp, which would have been submerged had we stopped here last night. It is probably perennial, as marks on nearby rocks indicate that the Indians have ground their corn there. Under a shelf at the mouth of a canyon [Whitmore Canyon] at which we halted this morning was a board bearing a number of carved names, dating from 1917 to 1923. One of the names was Edna Gass. The names are probably those of Mormon campers who came down the canyon from the plateau.

LINT: Rained at noon and after we ate lunch the wind blew so hard that Burchard was unable to use the alidade so we lay up for a while. Had several showers this afternoon. Dodge said that these people whose names were on the plank were homesteaders back from the rim on top. When Emery was rolling up his bed this morning he discovered that a 2½-inch scorpion had shared his sleeping quarters with him.

September 24, 1923 — 14,200 ft³/s

STABLER: We encountered a succession of side canyons today that had to be surveyed. This section is much broken by faults—hence the canyons in the broken walls [Hurricane fault zone]. We made about 5 miles progress by splitting into two parties—Birdseye, Moore, and Freeman surveying two side canyons while Burchard,

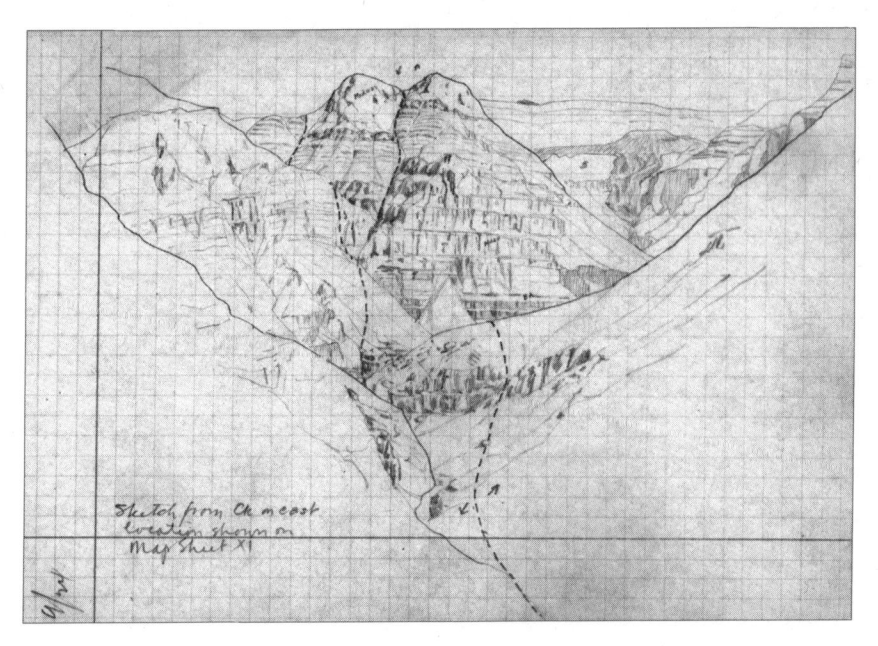

Sketch map of an unnamed canyon on the Hurricane fault zone at about mile 192 in western Grand Canyon.

R. C. Moore diary, p. 92. Courtesy of the University Archives, Spencer Research Library, University of Kansas Libraries.

Dodge, and I carried the main traverse. Our old map seems to be at fault in the location of Andrus Wash. Also it would not lead one to suspect the side canyons we have met. It is somewhere near right as to the general course of the river. Camp on a big sand bar with a pool of settled water, lots of bed ground, and driftwood [about 194 Mile Canyon]. The evenings are rather cool now so a driftwood fire and songfest after supper ended the day. Lava still continues. That near camp apparently came down the river. The source of much lava is cinder cones to the north seen when we climbed out Prospect Creek No. 3 yesterday.

BLAKE: Many riffles and side canyons were encountered today. It was also very hot. In several places the river is rapidly eating away the high sand bars which the recent flood deposited. Black willows are beginning to appear along the banks of the river that afford shade which is very welcome now that there are no high walls close to the river to answer the purpose. While sitting still in our boats yesterday several of us had the opportunity to see a blue heron at close range. The big bird sat upon a sandbar within ten feet of the boats and stood there for several minutes, but got frightened and flew away when I tried to get my camera so as to get a snap shot of him. We camped on a wide sandbar where there was plenty of wood. Camp was not made until late and supper was eaten at dusk. We built a big camp fire after dark and spent the evening singing all the songs we could think of.

FREEMAN: We saw our first flock of quail today. They were amazingly tame.

September 25, 1923 — 13,000 ft³/s

BLAKE: Although the river only dropped half a foot last night some of the boats were stuck in the mud this morning. Freeman's boat was stuck the tightest, and as he had loaded it up before trying to push off, it had to be partly unloaded before it could be extricated. Of the four boatmen Freeman was the only one who did not get into the mud and water in helping get the boat off.

STABLER: Lots of side canyons and little progress today ... Again we saved much time by running two parties, Burchard, Dodge, and I taking the river traverse and running two side canyons. The canyon is not so broken down as the previous day and we got some high straight walls in the afternoon. We seem to be leaving the faulted section and should have fewer side canyons to run hereafter. On reaching our camping place La Rue discovered a rattlesnake (and killed it) that was 3 feet 8 inches long and had 11 rattles. Quite a bird. Saw a few ducks today—the first for a few days. Double parties on the survey again tomorrow.

BIRDSEYE: Reached a deep canyon on the right (Andrus Wash?) at mile 136.8 [Parashant Wash]. I traversed up to 1692 at night and the next day completed the work up to 2100 about 6 miles from the river. Four and ½ miles from the mouth I traversed up a side canyon from the southwest but found it blocked by a 60-foot falls about 500 feet from the main canyon. The entire canyon is dry except for stagnant pools of rain water. Camped at the mouth of this canyon after a day's run of 3.8 miles.

September 26, 1923 — 17,300 ft³/s

STABLER: Birdseye, La Rue, Moore, and Freeman stayed to survey the side canyon at our camp and the rest of us went ahead with the river traverse and had gone about 5 miles before they caught up with us in the late afternoon. We are camped on a sandy flat at the mouth of a little clear-water stream that enters from the right [Spring Canyon]. This stream is both a joy and a sorrow. It furnished good water for drinking and washing—but it has to be surveyed probably for 2 or 3 miles. We shall split up again today in order to save time. The country downstream looks broken up as though it were the opening of a narrow valley instead of the deep canyon we have been following. The old map is hard to follow but it looks as though we were now about 20 miles above Diamond Creek. We are due there tomorrow but will be a few days late because of our delay at Lava Falls. Camp fire and songs tonight. It is cool enough to make a camp fire feel good nearly every evening though it is pretty hot in the sun at noon.

BIRDSEYE: Springs [in Spring Canyon] about ¼ mile up run about ³⁄₁₀ of a sec. foot but the canyon is dry above. Below the springs the canyon is filled with willows and arrow weed. River rose 3 feet during the night.

September 27, 1923 — 18,900 ft³/s

BIRDSEYE: I traversed up Spring Canyon to elevation 1960 where the box canyon is only 3 feet wide at the bottom and is blocked by several sheer falls with large boulders wedged in the canyon walls. Several pools of stagnant water below the falls but no running water above the springs. I traversed up a deep canyon from the right to elevation 2012 at mile 145.0 La Rue found several old Indian cliff dwellings about ⅓ mile up on the right with remnants of pottery, baskets and arrowheads, so named it Indian Canyon. Camped at the mouth of a deep canyon from the right with two deep canyons opposite after a day's run of 4.3 miles. At this point the river widens considerably and granite outcrops in the side canyons together with the wide river valley with many willow trees along the left bank led us to call the place "Granite Park." This night was the coldest on our trip so far, the temperature at 6 A.M. the next day being 52° Fahr.

LINT: Three quarters of a mile below camp Burchard worked 1½ miles up a side canyon on the south side of the river [205 Mile Creek]. At the mouth of this canyon is Rapid No. 46 with a fall of 9 feet [205 Mile Rapid]. There are several boulders in it but it has a clear channel in the center. I was the only one carrying passengers through; the rest of them walking. I took considerable water from the large back-lashing waves. Ran three more side canyons in the afternoon—two on the south and one on the north side.

STABLER: We are still in a much faulted region and I am absorbing a little geologic savvy though much of this region is too complex for me to figure out. Pretty cold last night and I wished for a blanket. May get one at Diamond Creek which we should reach in a few days.

September 28, 1923 — 18,000 ft³/s

LINT: Burchard, Stabler, and Dodge worked 2½ miles up the side canyon near camp and then crossed the river and worked the short side canyon opposite camp. The Colonel, Emery, and I worked the large side canyon on the south side of the river below camp. It had five different forks so we worked about 8 miles in all without lunch or water and got back to camp at 4.30. Freeman scratched his elbow yesterday and Moore has a weak tendon in his right ankle so they were unable to do anything today. It must be great to be a cripple—sometimes. Same camp as last night.

FREEMAN: I accompanied Moore, who geologized up [Granite Park Canyon]. We found here evidence of an old stock trail, the fairly recent tracks of a man and a fragment of tin can. Apparently the canyon can be ascended to the plateau. This is the day the packtrain was to have met us at the mouth of Diamond Creek. We will be at least two or three days late from present prospects.

STABLER: This creek survey work is hard on shoes and I finished Leigh's boots today. Am now wearing a pair of shoes Birdseye gave me. They are No. 10, but a pair of insoles and big wool sox make them fit. I will cut Leigh's boots down to make a pair of leggins out of the tops. Grub is running short in spots. No butter, beans, prunes, etc., but plenty of bacon, ham, flour, etc., so we have plenty to eat though are beginning to get less variety.

September 29, 1923 — 15,700 ft³/s

BIRDSEYE: Ran riffle #46-d just below Granite Park [209 Mile Rapid]. This is a long riffle around a large boulder island and has a fall of 18 feet from mile 147.2 to mile 148.6. The right hand channel around the island has the roughest water but either channel is good at this stage of water. We ran the left channel.[42] I traversed up a deep canyon from the right at mile 149.8 to elevation 1922 where it is blocked by a 100 foot falls [Fall Canyon]. This canyon is dry. At mile 151.1 I carried the rod up a small steep canyon from the right

---

42. 209 Mile Rapid is a lot different now than in 1923. The left channel around the island periodically has been closed by debris flows and opened by river floods in the late 1990s and early twenty-first century; few would go there except in kayaks or small inflatables and only at discharges higher than what the 1923 expedition was on. On the right channel, a block fell in 1978 from an alluvial bank into the river, creating one of the largest holes in a Grand Canyon rapid. Webb, *Grand Canyon*, 170.

to elevation 1734 where the canyon is blocked by a series of sheer cliffs [probably opposite Pumpkin Springs]. Burchard could see the entire length of the canyon from his instrument station on the river bank. Found several rock mounds which looked like Indian mounds but the initials R. P. nicely chiseled in the rock near these mounds indicate a previous visit by a white man. I traversed up a small steep canyon from the left to elevation 1745 at mile 152.0. At this point the canyon was blocked by sheer cliffs. I traversed up a deep canyon on the left to elevation 2031. This canyon is dry and climbs rapidly over many small falls. Camped on sand terraces on right bank at mile 152.7 after a day's run of 5.5 miles.

STABLER: I worked along the river with Burchard all day until late afternoon when we climbed about 600 feet and sat up on a cliff in the Bright Angel shale that we call Pa Snuff, it being the highest of a series of similar snuff-colored cliffs. We got down a little before dark to camp which had been made just below us. The walls were unusually beautiful today, spires and domes appearing at intervals to vary the monotony of sheer cliffs. Granite came up for a few hundred feet and then dropped under the Tapeats which continues with us. We should strike the Diamond Creek granite gorge today for we are probably only 10 miles from there.

September 30, 1923 — 14,600 ft³/s

LINT: The Colonel worked two side canyons today and Burchard three small ones. About 1½ miles below last night's camp and on the north side of the river I climbed up on top of the Tapeats to give Burchard a shot. While there I found two built up piles of small white pebbles. They were shaped like the cone of a volcano and the rim was about two feet high. The inside of them was perfectly level and covered with charcoal which was in very small pieces. They were about 18 feet in diameter and looked very much like a huge plate except that the outside surface sloped outward. These must have been used by the Indians for some of their celebrations.[43] There is an old well defined trail close to them but it hasn't been used for years. At Mile 155.8 below the mouth of the Little Colorado River is Rapid No. 47 with 9 feet fall [217 Mile Rapid]. Large waves on the north side and rocks on the south side at the head. Ran it by entering the "V" on the north side and cutting through to the south.

---

43. Lint is describing an agave roasting pit, used by several Native American groups (in this case, Southern Paiute) to cook agaves harvested just before they flowered. The cooking rendered the complex carbohydrates into sugars. Agaves roasted in this fashion constituted a large part of the Native American diet.

None of us took water except Freeman. Just below No. 47 are two riffles, the first one with a fall of 3 feet and the second 2 feet.

BIRDSEYE: At mile 153.2 Burchard traversed up a deep canyon from the left. At mile 153.9 I traversed up a deep canyon from the left to elevation 2100. We called this Three Springs Canyon on account of 3 flowing springs, one being within 200 yards of the river.

FREEMAN: The riffle around the bar above camp drove hard against the left-hand cliff, with an unexpectedly heavy boil on the right [probably Trail Canyon Rapid]. A horse-fly alighted on my forehead just as I put into this, and I had to let him bite until I came out at the foot. Result—big welt on forehead and a smeared fly.

STABLER: We are making every effort to get to Diamond Creek where mail and supplies are waiting for us so we put in a long hard day today and in spite of our side canyons we made about 5 miles. We camped at the mouth of one little canyon [220 Mile Canyon] and two more are in sight, so progress tomorrow is sure to be slow. Had a few mosquitoes around last night and one fed on me. I don't think, even were he an *Anopheles,* that there is much chance of his carrying malaria from some other human.[44]

October 1, 1923 — 14,400 ft³/s

FREEMAN: It is clear this morning, with apparently a slight rise in the river. Two more large side canyons inside of three miles gave Birdseye a full day, with Moore recording and I rodding. The first opened up to a valley above, and we went to the 2100-foot level on two forks almost on a uniform grade [220 Mile Canyon]. We were from 8:30 to 2 in this canyon. Below the bar of the latter there is a broad, shallow riffle filling the whole river, with a slightly sharper pitch at the lower end. It is 400 yards long and proved to have a fall of 10 feet [Granite Springs Rapid]. It offered no difficulty in running. A mile, with a couple of minor riffles, took us to the mouth of another side canyon from the left [222 Mile Canyon], which occupied Birdseye until too late to push on to Diamond Creek with the *Grand* as he had planned.

STABLER: This morning was a hard one. Burchard and I undertook to survey two creeks and adjacent territory and it took lots of hiking and climbing. We got back to the river after 2 P.M., both

---

44. In 1923, the U.S. Public Health Service was actively engaged in efforts to control malaria within the United States. The disease was not considered eradicated from the U.S. until 1951. Centers for Disease Control and Prevention, http://www.cdc.gov/malaria/history/index.htm (accessed June 13, 2006).

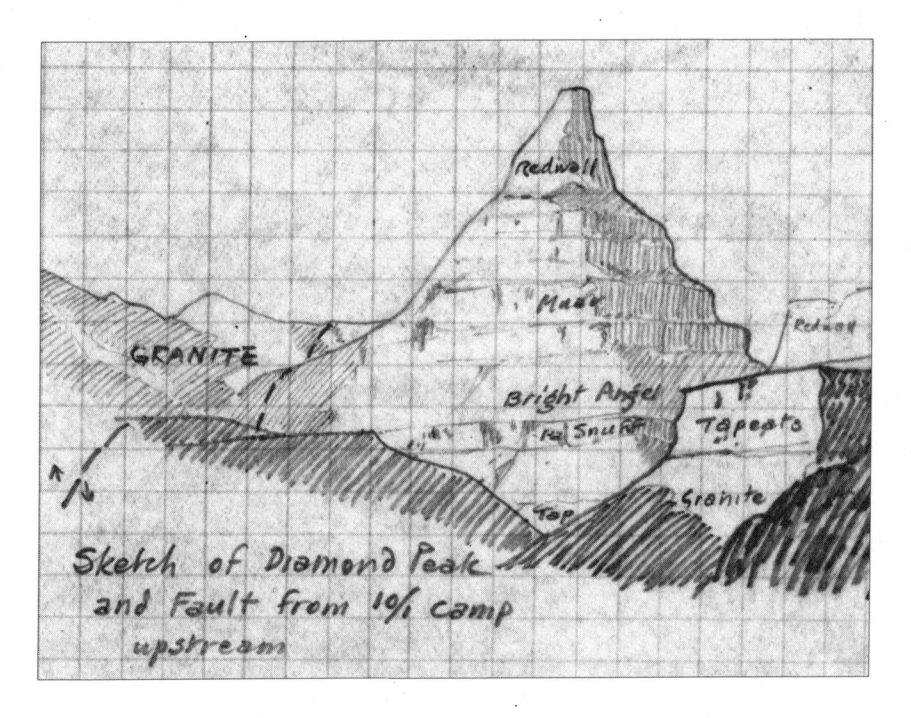

Sketch map of Diamond Peak in western Grand Canyon showing geologic formations. In all likelihood, this sketch was made from the camp at mile 222, river left.

R. C. Moore diary, p. 106. Courtesy of the University Archives, Spencer Research Library, University of Kansas Libraries.

---

pretty well tired out. Up one creek was an old camp with corral, etc., which we saw from above but did not inspect closely. On a point where we set up were two little rifle pits in the rocks—probably all the work of prospectors. Up the other canyon we can upon a bunch of burros. They were tame and doubtless belonged to the deserted camp we saw. We are camped in sight of Diamond Peak, quite a striking sight directly down the river [about mile 222].

BIRDSEYE: At mile 158.7 I traversed up a deep canyon from the left to elevation 2054. Found a small spring in granite gorge of the main fork at elevation 1940 and called this canyon Granite Spring Canyon.

BLAKE: A mile and a half below camp I saw what looked like a cave upon the side of the cliff and investigated it, finding it to have been inhabited at some time, as there was a tumbled down wall in front of the opening and smoke stains on the ceiling. I could stand upright in the entrance which was about eight feet wide. The depth of the cave was eighteen or twenty feet, the floor being level, while the roof sloped down to within four feet of the floor. Camp was made two and six-tenths miles below last night's camp, at the mouth of a small canyon. We still see at intervals, the remains of the lava which at one time flowed down the river channel but which has been eroded away except for an occasional bit which the river and weather have not disintegrated and carried away.

October 2, 1923 — 14,100 ft^3/s

BIRDSEYE: The *Marble* with Kolb, Koms and I and the *Grand* with Freeman, La Rue and Moore ran on ahead to mouth of Diamond Creek, reaching there at 9 A.M. Burchard continued the river traverse and reached Diamond Creek at 4 P.M. B.M. H-9 at mouth of Diamond Creek (elevation 1362.166) was reached at mile 163.8 and checked on with an error of 8.9 feet. Had we not adjusted our elevations by a -3.6 ft. correction at mouth of Havasu Creek to Evan's vertical angle B.M. we would have checked at Diamond Creek with an error of -5.3 feet in the line from Bass Trail crossing—a distance of 118 miles. As it is the error is -8.9 feet in 69 miles or 0.13 ft. per mile, which is good work considering the long stadia sights with many vertical angles. Found mouth of Diamond Creek at mile 163.9 or 68.9 miles below the mouth of Havasu Creek and 224.4 miles from the mouth of the Paria River. Found Roger Birdseye and Chas. Fisk had been waiting since September 28. They brought mail, supplies and newspapers with wildly exaggerated accounts of our possible disaster in the big flood. Records from the Bright Angel gauging station showed the rise to be 21 feet, while we measured approximately 20 feet at Lava Falls. One of the reports stated that an overturned boat with letters USGS had been seen below Boulder Canyon. So far as I know no such lettering is on any boat on this part of the river. The Water Resources Branch boat at Bright Angel broke loose and was carried away, but the description of the boat sighted does not tally with that of this boat, so we assume that the letters were USRS, and that the boat belonged to the Reclamation Service at Boulder Canyon. Roger and Fisk had brought the supplies in by wagon to a point 10 miles from the mouth of Diamond Creek where the old road was completely washed out. From that point they packed the supplies down on one saddle horse, making three trips. They had kept that horse at the mouth of Diamond Creek so we immediately sent Fisk up to Peach Springs with telegrams to the families of each member of the party. Also sent telegrams to Kingman, Los Angeles and Washington reporting our safe arrival. The telegram to Los Angeles was broadcasted by radio from the *Los Angeles Times* and we heard it read at 8 P.M.[45] Roger brought back our radio set with new batteries and the broken connection repaired and we found it worked fine in the narrow gorge at Diamond Creek. Camped at the mouth of Diamond Creek after a day's run of 3.7 miles and prepared to remain several days to make a large scale dam site survey supplementing that already done by Girand.

BLAKE: One rapid [224 Mile Rapid] was run, which we ran head on, or prow first, in order to get more kick out of it. Dodge stood upon my boat, holding to the rope to steady himself, and near the end of the rapid let go of the rope with one hand, thinking the worst waves had been passed, but a side wave took him by surprise, making him lose his balance and causing him to fall into the river. About 3 o'clock we began to think we might have to lay out over night as a big side canyon was sighted in the distance, but when we reached it saw the smoke of the camp at Diamond Creek, a thousand feet below on the opposite shore.

FREEMAN: The flood—considerably higher than the spring high-water of this year (112,000 second feet) —was taken advantage of by the Arizona papers to send out a lot of sensational screeds, all calculated to foster the belief that the party was in grave danger. The heading in a late paper read: "HOPE NOT GIVEN UP FOR RIVER PARTY!" Roger, acting with his usual good judgment, had run down the Kingman correspondent responsible for the most lurid tales and frightened him into repentance, and had also written reassuring notes to the families of all members of the expedition.

KOLB, LETTER:[46] Dear Blanche, I was afraid there would be some scare about the flood, but didn't think it would go to the extent it appears to have gone. We had rather an exciting night however ... No other incidents of importance have occurred. You may pay for this movie print and developing. I don't know what to say about you coming to Needles as I don't want you to drive there. Come by train if you like, but otherwise I will come home by train and see you both which may be better. Please write to mother and tell them

---

45. The paper ran a multi-part article, reporting from Peach Springs, with quotes from Moore's mother, Mrs. B. H. Moore, and La Rue's wife, Mabel. "Grand Canyon Party Dodges High Waters," *Los Angeles Times*, October 3, 1923 (accessed March 1, 2006).
46. Emery Kolb to Blanche Kolb, October 2, 1923, box 5, folder 628, Kolb Collection.

we are all ok. The only suffering we have done is overeating. With very much love to all, Dady

BIRDSEYE, TELEGRAM TO USGS:[47] Arrived Diamond Creek today no trouble except three days delay at Lava Falls forty five miles above Diamond Creek due to twenty foot rise in river beginning night September eighteen and reaching peak following night all men well and boats uninjured expect to reach Needles October Fifteen send twenty five transportation requests and twelve bills of lading also other mail to Needles marking all mail hold until called for.

October 3, 1923 — 13,900 ft³/s

SECRETARY OF THE INTERIOR HUBERT WORK, TELEGRAM TO BIRDSEYE:[48] My warmest congratulations to you and party. Stop I was very uneasy.

LINT: Lay around camp and rested today. Felix made some doughnuts and cinnamon rolls for dinner and light bread for supper. Charley came back at 5:30 and brought us in some grub as we were running low. Emery was quite sick last night and lay in bed all day.

STABLER: This morning La Rue, Moore and I climbed out on the Tonto rim west of camp and looked over the possibilities for a high dam at the Girand site. Girand has picked the best site for a dam up to 250 feet in height but for a dam to back to Supai—440 foot—it is scarcely suitable. For such a dam the best location is below Diamond Creek because the granite runs somewhat higher and also because there is less chance for trouble from faults which are somewhat numerous in this locality. For such a location spill would be from Diamond Creek via a gulch to the Colorado some distance below the dam site. The lowest point in the saddle would be 500 feet higher than the crest of the dam so that a tunnel spillway 1000 to 1500 would be required. A location almost as good would be a few hundred feet above the Girand site. Spill would then be by a similar tunnel to Diamond Creek. Subject to foundation conditions, which are speculative, the lower site seems preferable. We had quite a hike, taking a light lunch and small canteen of water along and found them useful. We circled around and came down a gulch half a mile up Diamond Creek. The view up Diamond Creek, up the river, and down the river, is simply wonderful, quite equal to anything up at El Tovar. If it were made accessible so that tourists

---

47. Claude Birdseye to Glenn Smith, October 2, 1923, Birdseye Personnel File.
48. Hubert Work to Claude Birdseye, October 3, 1923, Birdseye Personnel File.

could go out on the rims it should make a very popular stopping place. This was the first tourist point on the canyon but parties had to come by stage coach down a very difficult road that was continually washing out and in addition never had a chance to get the views from the rims.[49] Took a good bath and washed up all dirty clothes this afternoon. Both operations were needed. Diamond Creek is a nice little clear water stream and afforded the opportunity.

October 4, 1923 — 13,800 ft³/s

FREEMAN: A poker game lasting until 11 tonight proved the worst dissipation of the trip. R. and C. Birdseye principal contributors.

STABLER: Early this morning, Burchard, Dodge and I started out on what we thought was a two-day survey of Diamond Creek and Peach Springs Wash. We took a small canteen and light lunch along. We completed the work on Diamond Creek soon after noon and ate our lunch in the shade of a rock. We then decided to tackle Peach Springs Wash which leaves Diamond Creek about a mile from camp. This we managed to finish also and make our way back to camp just at dark. We ran 5 miles of survey so had 10 miles of hike including a lot of rough travel and quite a little climbing so we were pretty tired and ready for bed after supper. Meantime Birdseye had started a dam-site survey at the mouth of Diamond Creek which he will probably complete tomorrow. We found a tarantula on our way back from Peach Springs Wash, the first seen by the party. As far as we went Diamond Creek maintains its character as a clear-water stream though some seeps had a slightly sulphurous odor.

KOLB, LETTER :[50] Dearest girl [Blanche], I was sick night before last but better today. Too much corn after not having any. We will likely leave here Sat. or Sunday. These long stops without much action sort of gets me. I am beginning to want to see the end come but it is just a little over two weeks more. Don't you think it best that I come right home a day or so to see Edith even if I do go with Johnson a few days.[51] Please order from Eastman Kodak Co. Rochester N.Y. six hundred feet perforated negative film, cut in one hundred ft. lengths sent at once C.O.D. Grand Canyon, Ariz. Can't think of anything else now. I sayed [*sic*] in bed yesterday and

---

49. About 1883, Julius H. and Cecilia M. Farlee opened a small, rustic hotel along Diamond Creek less than a mile from the Colorado River. Tourists came in by stage coach from nearby Peach Springs. A modest business at best, the Farlee Hotel completely closed by 1901. Anderson, *Living at the Edge*, 38–39.
50. Emery Kolb to Blanche Kolb, October 4, 1923, box 5, folder 627, Kolb Collection.
51. Johnson, possibly a friend or a client, wanted Emery to join him on a hunting trip.

was too weak to write or think about it. Felix fixed me up with some fine milk toast last night. Get all the clippings you can about the flood etc. Love to both, Dady.

October 5, 1923 — 13,500 ft³/s

FREEMAN: At Birdseye's request I wrote an account of the flood that overtook us at Lava Falls, to be sent out to the *Times* radio and as a news dispatch. I have made it a 2,000-word account of the voyage from Hermit Creek to here, but mostly devoted to the flood.

LINT: The Colonel worked on the dam site all day. I recorded for him and Dodge and Blake rodded until noon and then only one rod was needed so Blake stayed in camp. Five Walapai [Hualapai] Indians rode down the creek this morning, took a look around, and went back.

STABLER: Around camp all day checking notes, helping Moore get geologic data from the notes I had taken, washing clothes, repairing shoes, etc. Burchard completes his sheets to send out with the last mail in the morning.

KOLB, LETTER:[52] Dear Blanche, Though we remain until tomorrow, Roger has no horse feed and leaves today. Dam sites are being surveyed. I have my check for August. Hardly like the idea of sending it signed through the mails. Freeman is sending a letter concerning the flood etc. to K.H.J. This probably will appear in the *Times* paper, also. I am a little dubious as to an early landing at Needles. We have now been here four days. Felix the cook is feeding us sumptuously. Doughnuts, chocolate cake, etc. Last night we had some of the finest corn fritters you ever ate. I am sitting in bed facing a rapid on my left and broken walls on my right, towering 2500 ft. Leigh on one side, Blake on the other, not yet awake. The sun is just tipping the cliff. It is really more picturesque at this point than I thought it was. I repaired one of the engineers shoes yesterday and did some washing. We have just about thirty-five miles work yet but will then take several days to get to Needles. If it were not for Johnson I would go on to the gulf. If you don't come to Needles keep me posted about his trip. It will only be a little while until Felix will be saying "Come and get it" so presume I had better dress. Don't worry or let any scare head newspaper talk worry you. We still have all our boats intact and likely will be to Needles. I haven't written Edith from here and won't get a chance to so you give her my love. I hope she gets along at school ok. She must not attend too many dances. With much love to both, Dady.

---

52. Emery Kolb to Blanche Kolb, October 5, 1923, box 5, folder 626, Kolb Collection.

7

Feeling Their Oats

Diamond Creek to Needles

They faced down a flood and managed to pull through without losing any equipment. Their head boatman flipped his boat in what until then was an unnamed rapid. Their original cook, who had eye problems but also could not stand the incessant squabbling among the crew, left at Havasu Creek and was replaced by a man who was illiterate and spoke poor English. In some ways, Felix Kominsky would become the iconic image of the trip, and he brought a sunny disposition into an often volatile personality mix. The 1923 USGS expedition lost valuable time during the flood at Lava Falls, but they made some of it up by their professionalism—they could work together even if some of the crew obviously did not like each other.

Now, most rowing river trips end at Diamond Creek, because Lake Mead begins about 15 river miles downstream. Shorter trips launch here; once these trips looked forward to solitude in the less-traveled western Grand Canyon, but now this reach is jammed with boat traffic and tourists. In some ways, the 1923 USGS expedition was beginning at about this point as well. They faced two fearsome rapids ahead—Separation and Lava Cliff—that Powell had built up in the public imagination for their potential hazard. Although head boatman Kolb had seen these before, he was now a little shaken by his mishap in Upset Rapid, leaving the two junior boatmen to question his judgment. One, Lint, would find his voice as a diarist, and his drawings document those large rapids now beneath the surface and sediments of Lake Mead. Freeman, whose boating judgment was repeatedly questioned, had the flip that everyone expected but with minimal consequences.

Travertine Falls at mile 230.5 in western Grand Canyon, 1998.

Dominic Oldershaw photograph, Stake 1919, Desert Laboratory Collection of Repeat Photography.

The fractious trip splintered still more. Once again, it was left to the gray-haired topographer, Birdseye, to pull his expedition together. The heavy work took its toll on the boatmen, who were repeatedly hurt carrying equipment over the rocks. Birdseye began to express his admiration for his fellow surveyors, particularly Burchard, who worked through pain to finish the heroic surveying that they had begun more than two months earlier. The river would help, as the summer rainfall season was over and the crew would face steady flow for the remainder of their journey.

October 6, 1923 — 12,900 ft³/s

LINT: The Colonel, Dodge, and I worked on the dam site [Diamond Creek] until almost noon. Spent the rest of the day in camp. Roger left for Peach Springs at about 9 o'clock this morning. Emery and I have had the radio up for three nights and we have listened in on some fine programs.

BLAKE: Moore, La Rue and Freeman hiked up Diamond Creek this morning. I painted the new oars and my old oars with linseed oil. Emery and Freeman also oiled their oars. We played five hundred then trimmed each others hair. The wind blew up stream all afternoon, filling everything with sand. We played cards until after nine then Leigh and I listened to the radio.

FREEMAN: Taking my large camera, I set off at 9:30 to walk to the source of Diamond Creek. The canyon boxed at the end of three miles, ran on very narrow for a mile and a half, and then opened out again. The gradient except for a couple of falls in the boxed section, was almost uniform. I passed several striking pinnacles, and here and there, through narrow side canyons, fine vistas of the higher formations, red and yellow in the sunlight. I found what appeared to be the source of the stream at the end of seven or eight miles, just where it reached the Tapeats and the canyon broadened out into a valley, a quarter of a mile wide, running on to the base of the outer sandstone and limestone cliffs. About a third of the stream came from a big spring in a thicket of willow and rushes, and the rest from the gravel bed of the valley under which it seems to flow. Probably it appears again above. It was nearly dark when I returned to camp.

October 7, 1923 — 13,000 ft³/s

LINT: Diamond Creek Rapid (No. 50) starts at the mouth of the creek and extends for 0.7 of a mile. The first section of it has a fall of 10 feet, the second 5.5 feet, and the third 6 feet—a total of 21.5

in all. The first section has three rocks in it in the shape of a V with the point upstream. We ran this part of it by keeping close to the west shore all the way through. Freeman was the last one through and for some unknown reason (?) he got too close to the shore and hit a rock, stern on, and knocked the iron hand hold through the stern. He repaired it about half a mile down stream. The second section is a short drop and at the end of it is a bad bunch of huge boil-ups and whirls. Entered it on the west side. The whirls sucked my boat down so that I took water over the gunwale.[1] The last section is just a stretch of swift water with some large waves. About a mile below Diamond Creek the marks on the cliff show that the flood of September 18-19-20 was 32 feet above the present stage of water. Rapid No. 51 is about 2.5 miles below Diamond Creek and it has a 10 feet fall [Travertine Rapid]. No rocks but lots of large waves which are bad for there is a fairly sharp bend in the rapid. The Colonel worked up a side canyon on the south side of the river from this rapid [Travertine Canyon]. While working his side canyon the Colonel dropped his instrument and it rolled about 15 feet down the slope but all the damage done was that the rule was slightly hurt. Blake forgot and left his life preserver on the bar at No. 51. Camped on the south side of the river near a small stream that fell over a 200-foot cliff about a hundred feet from the river [Travertine Falls].

STABLER: Rather an eventful day. As usual recently I rode in the *Boulder* with Leigh at the oars and Burchard on the stern. In landing below Diamond Creek rapid Leigh scraped a sheet copper plate loose on the bottom of the boat. We soon came to a creek that had to be run out and got the work done without trouble. Burchard and I were coming back together, I ahead, and had nearly reached the river when I heard a great commotion behind. Looking back I saw that Burchard had slipped at the edge of a pool of water and landed plunk in the middle of it, sitting down with only head and feet out. He was carrying the plane table on tripod over his shoulder and it bumped one corner on a rock but was not injured, nor was the map on it even splashed. Burchard was well soaked and also got a rather bad bruise on the lower ribs and thought, mistakenly, that he had a broken rib. He was able to keep on working the rest of the day. He surely was a comical sight in the water.

---

1. The whirlpools and "swirlies" of the gorges, both Upper and Lower Granite Gorge, are well known among boatman for their unpredictability and potentially disastrous consequences.

BURCHARD:[2] I didn't know I was hurt. Birdseye had skipped a little canyon so I went in to get it—I slipped into a little pool. I fell against corner of plane table. I knew I got punched in ribs, slid 10' and hit table. I went on working—that night it began to hurt. Stabler knew what happened. I wanted to finish job. Next morning I was in real pain ... I told Stabler I would have to give up—Stabler said you probably have broken rib—my head was spinning. Stabler called Birdseye to come in—Birdseye said that's a busted rib—wrapped me around & around with 15' of tape. I never felt another thing. I finished the trip with that tape.

FREEMAN: Earlier in the day he [Burchard] sat down upon a barrel cactus, but was saved from painful injuries by the leathern patch on the seat of his pants.

BIRDSEYE: With the exception of the first day's work below Lee's Ferry Burchard had made the entire river survey and wanted to complete the job and join his old work just above the mouth of the Grand Canyon. This would make him responsible for the entire survey of Marble, Grand, Boulder, Black, canyons from a point 7 miles below Lee's Ferry to the Bulls Head reservoir site about 40 miles above Needles. The fractured rib caused considerable pain but strips of adhesive and elastic bandages made it possible for him to continue his work.

BLAKE: At Garnet point, which we passed during the afternoon, a stream of water flowed down over the granite in a succession of falls [Travertine Grotto, in Travertine Canyon]. Two hundred yards from the mouth of this stream its narrow canyon is tunnel like in its darkness. A water fall about seventy-five feet high tumbles and slides down over the slick rock. A large boulder has lodged in the canyon about a hundred feet overhead, forming a bridge across the narrow chasm. The river has been swift all day with one small rapid or large riffle following another. Even with the swift current to aid the progress of the boats, rowing has been hard on account of the upstream wind. Camp was made near a stream of water which pours over a cliff two hundred feet above the river, and which came tumbling down its self made slide, the water having built a limestone ridge ten or twelve feet thick by depositing some of its lime content, making a smooth moss covered course, down which the silvery threads of its stream slide noiselessly. Some of the party bathed at the foot of this unique water fall but most of us dreaded too much

---

2. Handwritten notes, undated, apparently from an interview with Roland Burchard by Otis Marston, box 25, folder 17, Marston Collection.

the combination of cold water and cold wind to do more than regular washing of face, hands and arms.

October 8, 1923 — 12,300 ft³/s

LINT: We hit our first rapid for the day 1400 feet below camp. No. 52 with 12 feet fall [231 Mile Rapid]. Landed at the head of it and went down to look it over and while returning to the boat it started to rain. This rapid is in two sections, the first one having two submerged boulders in the south side of the "V." The lower section has three boulders, two submerged and one just barely sticking out of the water. I ran the first section on the north side and then cut through the tail of the first section and landed at the head of the second section. Burchard made a station on the point and I ran this part of it alone. I pulled around the partly submerged rock and then pulled in to the south shore and picked up my passengers. Blake followed the same channel but he didn't have to land as I did. Rapid No. 53 has 5.8 feet fall and is 6.7 miles below Diamond Creek [232 Mile Rapid]. It is a very narrow rapid with all the fall in a very short distance. It has one rock at the lower end on the north side and some large waves at the end of the "V." Emery and Freeman ran it stern on and took quite a little water. Blake and I hit it prow on and cut through to the south side thereby missing the big water. All the passengers walked along the north bank. Ate lunch just above Rapid No. 54 which has a fall of 9 feet. Landed at the head of it to look it over and also to land the passengers for they said that it was too cold to get wet. Blake, Emery, and I entered it in the "V" and then pulled in to the south bank. Freeman entered it stern first and rode all of the waves taking a considerable amount of water. There is a line of rocks extending about half way across the river in the head of the rapid. Had another shower which lasted about half an hour. The Colonel worked a side canyon on the south side of the river at Rapid No. 55. Camped on the bar made by this canyon. 4.6 Miles today with a fall of 46 feet.

BLAKE: A large rapid was run which had a few dangerous rocks in the lower end of it [231 Mile Rapid]. A short distance below there was another rapid [232 Mile Rapid] which was very short and steep, with a very sharp rock, alternating disappearing and protruding above surging waves.[3] Freeman came uncomfortably close to the

---

3. This rock is notorious among the few oar-boat travelers who regularly pass through the river canyons downstream from Diamond Creek. These are the "fang rocks," and the rapid is facetiously known to some as "Killer Fang Falls." Otis "Dock" Marston called it "Requiem Rapid," as he thought it to be the spot where the Glen and Bessie Hyde party

rock. Felix says, "What's a matta, I am so good a driver as he is." While another remarked that there must be a magnet on the stern of the *Grand*. Leigh and I followed, leaving our passengers to walk, as had the other boats, but contrary to the custom of going stern first we decided to pull hard prow first and so get more kick out of it. This we did and made a good run, taking practically no water, though the boats reared and pitched half out of the water. It being cold and windy today, it was decided to make coffee for lunch so I had to unscrew a hatch and give a can of coffee to Felix as the *Marble* was going ahead and have lunch ready at the mouth of a side canyon which could be seen in the distance. While stopped for lunch a measurement of the height of driftwood from the river was made. Some old drift was found to be fifty four feet above the present level of the river, while the drift deposited by the recent flood was only about thirty feet higher than the water level. After crossing the river after lunch, so that our passengers could walk around the rapid, we each made a good run. Only one of the boats took enough water to make bailing necessary. Showers fell during the afternoon. The river ran more smoothly this afternoon until a rapid was reached where it was necessary that everyone ride the boats [Bridge Canyon Rapid], which meant that all would have wet clothing. It being so late in the afternoon it was decided to camp at the head of the rapid rather than undergo the discomforts of wet clothing so late in the evening. We put up the radio and got Salt Lake, KYL, *Deseret News* broadcasting station and KHJ the *Times* station of Los Angeles, which we rely on for our best entertainment and news. There being no driftwood at camp we climbed the side hills for mesquite wood, and procured enough to enable us to have a large bonfire during the evening.

STABLER: The canyon has been a narrow granite gorge all day. Above the granite is a Tapeats cliff that has been at elevation 2150 or near that. Occasionally we could see beyond the Tapeats magnificent walls of the higher sedimentaries. This is the greatest granite gorge of the three.

October 9, 1923 — 12,400 ft³/s

STABLER: Soon after starting today we came to a remarkably fine dam site—great opposing granite walls 650 feet above the river, creeks and gulches offering spillway facilities etc [the Bridge

---

perished in 1928. Brad Dimock, river historian, also calls them the Fang Rocks but is somewhat ambivalent as to the Hydes' fate at this spot; Brad Dimock, *Sunk Without a Sound, The Tragic Colorado River Honeymoon of Glen and Bessie Hyde* (Flagstaff, Arizona: Fretwater Press, 2000), 257–65.

Canyon–Gneiss Canyon dam sites]. We surveyed it on the small scale as a possibility in connection with the Arizona schemes [proposed by James Girand]. After passing this the granite walls lowered somewhat so it appears to be the last site for a very high dam in the granite.

LINT: Ran No. 55 the first thing this morning and carried all the passengers as they would have had a hard climb to get past the cliffs [Bridge Canyon Rapid]. Pulled through the right side of the "V" and kept as close to the right shore as possible and yet miss the three rocks that were close in. Had a clear channel down the center but it was just a line of cross waves and we didn't want to get soaked so early in the morning. 12 feet fall in 1700 feet, most of it being in the head of the rapid. Freeman hit the first rock in the rapid but without damage to the boat. Rapid No. 56 with 9.8 feet fall is 10 miles below Diamond Creek [Gneiss Canyon Rapid]. It is a straightened out "S" rapid with rocks at the head on the left side, rocks on the right about half way down, and one large rock on the left side at the foot, and most of the current runs into this rock. Ran it by pulling through the left side of the "V" at the head, stern on, and then keeping in the center until almost to the rocks on the right. Turned around, prow first, and pulled as close to the rocks as possible and kept on pulling to the right in order to avoid the rock at the bottom and on the left. Blake followed the same course but shipped a little more water than I did. Emery and Freeman made it through OK without taking much water. There is a side canyon on each side of the river at this rapid and the Colonel worked both of them. Burchard worked a side canyon about half a mile below No. 56 and we ate lunch when he got back to the river. Just before lunch I accidentally cut my left wrist coming very close to the artery. The Colonel thought it best that I take it easy for awhile so Blake took Burchard and Dodge navigated my boat while I enjoyed the scenery. Rapid No. 57 [237 Mile Rapid] with 6.5 feet fall is 11.4 miles from Diamond Creek and is just a straight chute of water and big waves. Dodge and I were the last ones through and we hit it head on to get a little kick out of it. Made 4.5 miles today with a total fall of 43 feet and camped at the Rapid No. 58 which some of the fellows think is Separation Rapid.

FREEMAN: About 4 o'clock a deep roar indicated the approach to a bigger fall than anything we have recently encountered, and the opening of two canyons immediately opposite each other at the head of it appears to identify the rapid as Separation, where the first Powell party was deserted by three men. We had not expected to find this historic point for several miles yet—18 miles from Diamond, instead of the 13 we have covered so far. The nature of the walls, the two opposing canyons and the sequence of riffles fit closely the Powell and Stanton descriptions. The canyon on the right, at the mouth of which we have camped, appears to have just such a beach as Powell paced back and forth upon in coming to his momentous decision, while it also opens up to the north in such a way as to lead to the belief that it is a favorable route of egress from the main gorge [Powell's trip camped along the left bank, not the right]. Kolb inclines to the belief that the rapid is not Separation, principally because he misses a large mid-stream rock that is associated in his mind with that point. We shall doubtless know more about it in a day or two.

October 10, 1923 — 13,000 ft³/s

FREEMAN: After looking over the rapid again this morning, Kolb announced positively that the absence of the big rock in the middle convinced him that we had not yet come to Separation.[4] Whatever the rapid was, it had to be run. There was no great enthusiasm displayed by the passengers for riding, but Dodge's attempt to climb down with the rod along the right walls ended in failure. The left wall proved to be sheer, though only about 500 feet in height, rather than the thousand estimated by Stanton. The right wall was more broken and not so high as its vis-a-vis, but still offered no possible chance to get around more than the first riffle. It is probable that Powell portaged his boats over a portion of this part of the wall.

LINT: Separation Rapid, No. 58. Spent quite a bit of time looking the rapid over for the walls are so sheer that it would be impossible to walk around without a rope to get down over the walls. It has a group of boulders in the "V" on the right side, the side canyons which are opposite each other having washed these rocks in. Ran this part of it by keeping as close to the rocks as possible and then pulling to the right to avoid the two huge back lashing waves on the left side where there is a small bar. All of the current runs into this point so we all came close to it.

BLAKE: Leigh was determined to row his own boat thru the rapid this morning [despite his injured wrist], so Dodge came back to my boat. We were the first ones thru, Dodge standing in a crouching

---

4. This is at least the third instance where Kolb's memory of the locations of rapids or their characteristics proved to be incorrect. The crew very definitely faced Separation Rapid.

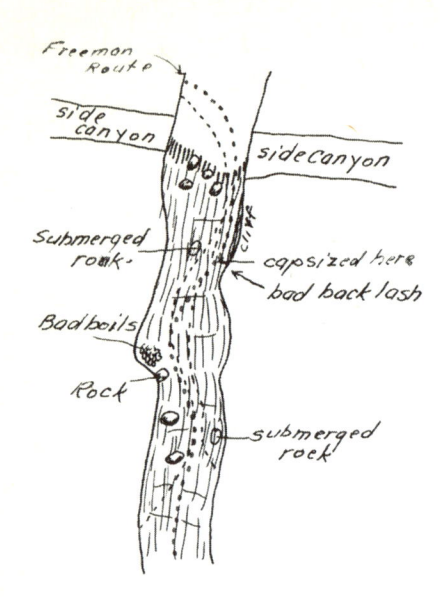

Sketch of Separation Rapid by Leigh Lint showing the route of Lewis Freeman's run and subsequent flip.

Lint diary, p. 31, courtesy of George Lint.

position on the stern, holding tightly to a life line.[5] The first few waves merely pitched the boat from side to side, slopping in a little water, but when about the middle of the first section of the rapid the boat plunged to the bottom of the trough of a large wave then reared its stern high in the air, but not high enough but what the upper few feet of the crest broke over Dodge with such force as to knock him half off the boat, pouring on over into the cockpit. We made a landing, however, above the second part of the descent, where Dodge held the rod for a turn.[6] Emery then came thru, and soon after Leigh's boat was seen pitching and bouncing to the quiet water below the first stage. We did not see Freeman make the start and so decided that La Rue was taking pictures or that Moore was geologizing which caused the delay.

FREEMAN, REPRISE: The *Glen* was the only boat I saw run, as I went back to the head of the rapid to take a picture of the two canyons and other features described by Powell. On being told that Kolb had carried dangerously near to the righthand cliff, I decided to take a chance at putting into the heavy water at the head of the second riffle, in order to keep well to the left and be in a better position

---

5. La Rue's movie footage shows Dodge standing on the deck of Blake's boat as it enters Separation Rapid. This clearly demonstrates how fearless Dodge was, in contrast to La Rue and his reaction to his ill-fated run through Separation Rapid.
6. Despite their concern about the whitewater, the survey was paramount. This act of surveying while boating through whitewater is now time-honored among certain USGS scientists, notably the authors.

to hold the boat away from the opposite wall. These waves, which I had not seen at closer range than 200 yards or more, proved a good deal larger than I had judged them to be. While holding the boat quartering to the main line of waves on the left, an unexpected and unaccountable comber from the right caught her broadside and threw her bottom-up in an instant. My only mental picture of the incident is of La Rue's legs spidering against the sky as he spilled off the stern hatch, all but falling into the cockpit. Then darkness and much rolling and tumbling of water. To keep from being cupped under the cockpit as Kolb had been in his upset at Havasu [actually Upset Rapid], I let go my oars and tried to drop down a few feet. In this the swirls gave me considerable help, for when I started back to the surface I found that it had become misplaced; or at least that I couldn't find it. The several succeeding seconds were as active as unpleasant. I pawed water and gulped water but didn't seem able to displace enough to bring the surface down where I could use it. Finally a violent swirl carried me to the top, but only to suck my protesting anatomy back so quickly that the breath of air I tried to get ended in a bite of water. This time the submergence was brief, and as my head came up again I managed to gulp a lung full of mixed air and spray. The downward pull was still strong though intermittent, and my head still ducked under now and then as I lunged off after the boat, floating ten or fifteen feet down-stream of me. I had the feeling that my saturated kapok jacket was giving very little support, but this may have been due to the drag of the current on my very heavy pair of hob-nailed shoes.

FREEMAN, DIARY: Pulling up to the boat, I found Moore, minus cap and glasses, riding comfortably and contentedly by one of the side ropes. La Rue was not in sight. I was about to essay some jocular remarks, when the sight of the right hand cliff apparently closing down above us told me that we were running full tilt in the middle of the current setting against that sheer granite wall, the one obstacle I had been most desirous to avoid. I muttered something to this effect. Moore nodded comprehendingly, as we exchanged of sickly grins. The white, untanned holes behind the places where the yellow shadows of Moore's amber glasses had fallen doubtless made his seem more solemn and owl-like than it really was. Moore could have been no less grotesque ... A rocking and jolting of the boat told when I was thrown back by the reflected wave from the wall ... When it [the boat] made up its mind to go on down the river according to the original plan, the worst of our immediate anxieties was over. Shipping swiftly and fairly smoothly on the edge of the back-thrown wave, it ran parallel to the wall for 20 or 30 feet before running beyond a jutting point into the choppy but innocuous water on the verge of the eddy fellow. Here a quavering call from the other side of the boat assured us La Rue was still a passenger, if not greatly enamored of the method of transportation. A splash and a quickening of speed signaled the entrance to the third riffle. From above, squarely in the middle of this, the current broke in a broad but not very high wave over what must be a large flat reef of rock, fairly well submerged at this water. It did not appear that there would be a violent collision if any, but it was reassuring to find that our derelict was giving the obstacle a berth of several feet on the left. The boat danced at a lively rate through the riffle, but not in a way to endanger our hold. Moore climbed up on the boat as the *Glen* pulled up and I followed. The rope along the bottom, put on the day we ran Hance Rapid, was a real help both in climbing up and in holding on. La Rue made no effort to climb up on the bottom of the *Grand*, but was helped over the side of the *Glen* by Dodge. I passed the painter of my boat to Dodge as soon as I could work along to where it was trailing in the water. Blake began pulling hard for the right wall, where a cove offered a chance of holding on. He made good headway, considering the drag of the boat and the current, but before he was half way in we passed a bend to the left which revealed the head of another rapid, with a huge rock about 25 feet high in the middle, against where the current was crashing in a great up boil. (Kolb told me later that this was the rock he had erroneously associated with Separation Rapid). It was plainly no place to bump into with an overturned boat. Blake made the cove for which he pulled with the *Glen*, but only Dodge good judgment in paying out the *Grand's* painter prevented him from being dragged out again. He caught a bight in a projecting knob of granite and held until we swung in against the wall. Here Moore and I were able to work up to a point where there was enough of a ledge upon which to place a foot and get a lift to right the boat. This was done with little trouble as soon as Kolb and the *Marble*, with Birdseye, had pulled in to help. Borrow a bucket from Kolb, I had my cockpit empty in a few minutes.

FREEMAN, REPRISE: Greatly to my relief, removal of the hatches revealed only a few inches of water in either hold and no damage to outfit, cameras or provisions. My worst loss was an oar; those of

Moore his glasses and a number of geological specimens left loose in the cockpit. We were dried out and ready to push on inside of an hour. Rather good luck throughout for a spill in so sloppy a corner. Stanton, who messed-up in almost the identical spot, smashed a boat and had a rather desperate time of it personally, according to his account.[7] For the benefit of those who had been overly wet during the morning, the Chief delved into the snake-bite corner of the medicine-chest.

BIRDSEYE: Moore and Freeman seemed to thoroughly enjoy the experience but La Rue had been sick for two days and his nerves were well shattered.

LA RUE:[8] The *Grand* with Moore, La Rue & Freeman upset in this rapid. Had a narrow escape. La Rue swallowed lots of water and came near be crushed between the upturned boat and the rocks in the wall.

MOORE:[9] I don't know that the tip-over in Separation was due to Freeman's boat. I was surprised that his broad beamed boat would tip over. I reached for Freeman. Freeman was no boatman but never lost his nerve.

LINT, REPRISE: Rapid No. 59 [241 Mile Rapid] with 8 feet fall is just below where Freeman pulled in to unload his boat. It has a rock at the head of it a little bit to the left of center and another one almost directly below it in the center of the big waves. The current runs directly for a huge boulder is on the right side about 400 feet below the submerged rock in the waves. Ran it by going to the left of the first boulder and dropping down through the left side of the big water and just missing the second rock. Had quite a pull keeping away from the whirl and the lower boulder. Ran No. 60 just before lunch. It has 8 feet fall and has two large back lashing waves. There is one submerged rock in the center of the "V" at the head and one protruding rock on the right just below. There is also a submerged rock on the left of the center which makes a big back-curling wave. Ran it by going to the right of the first rock and just missing the next one by going to the left and then pulling to the right to keep out of the big water. All of the passengers walked for they had a good slope to walk on and most of them seemed to like the idea of staying out of the rapid. Camped on the right side of the

---

7. Stanton did in fact have a similar experience in Separation Rapid in 1890; Smith and Crampton, *The Colorado River Survey*, 244–47.
8. La Rue, photographic diary, 41.
9. Otis Marston, interview with Raymond Cecil Moore at Lawrence, Kansas, May 1948, box 152, folder 29, Marston Collection.

Emery Kolb piloting the *Marble* through Separation Rapid, river mile 240, October 10. Kolb's passengers that day were Birdseye and Kominsky.

E. C. La Rue photograph, Topography D84, courtesy of the U.S. Geological Survey Photographic Library.

---

river 4.9 miles below last night's camp. A total fall of 61 feet today. P.S. Later we found that in the upset this morning they lost 3 caps, one water bucket, two geologist's picks, an oar, and La Rue lost a note-book with some photographic records in it.[10] Blake found the oar this afternoon.

---

10. This may be the reason why La Rue's remaining diaries end at Clear Creek—he may have lost his diary in Separation Rapid. However, as previously noted, we have his complete diary of photographs.

Sketch of Lewis Freeman and Frank Dodge by R. C. Moore.

Edith Kolb diary, Emery Kolb Collection, box 14, folder 1756, courtesy of the Cline Library, Northern Arizona University.

October 11, 1923 — 12,800 ft³/s

BIRDSEYE: After a short run on the morning of October 11 a party reached a large deep canyon coming in from the left [Spencer Canyon] opposite a low lava capped cliff. This canyon is known as Spencer Canyon or more probably Mattowitteki Canyon.[11] Julius Stone, one of the early river voyagers, named the rapids at the mouth of this canyon "Bold Escarpment Rapids" [Lava Cliff Rapid]. This name seems inappropriate as there is no real escarpment in the vicinity and members of the party could see no significance in the word bold. So the Indian name Mattowitteki Rapids was adopted [it did not stick]. This is one of the worst rapids on the river having a fall of 17 ft in a few hundred [feet]. The channel is dotted with rocks all the way down and there is apparently no safe way through. The Kolb Bros. portaged the upper part in 1911 [1912] but with a 6 ft lower water stage they had an easy landing in a cove at the brink of the fall. Records show that one of Powell's [boats] broke loose while lining and one man was thrown out but rescued. The other boat was then run through and caught the first. Stanton's record

---

11. Meriwhitica Canyon is a major tributary of Spencer Canyon, which in turn was the tributary that created Lava Cliff Rapid. Meriwhitica Canyon derives its name from a Hualapai term and means "hard dirt or hard ground"; Spencer Canyon is named for Charles Spencer, a prospector and ally of the Hualapai Tribe. Lava Cliff Rapid is now submerged beneath Lake Mead; Nancy Brian, *River to Rim* (Flagstaff, AZ: Earthquest Press, 1992), 128–29.

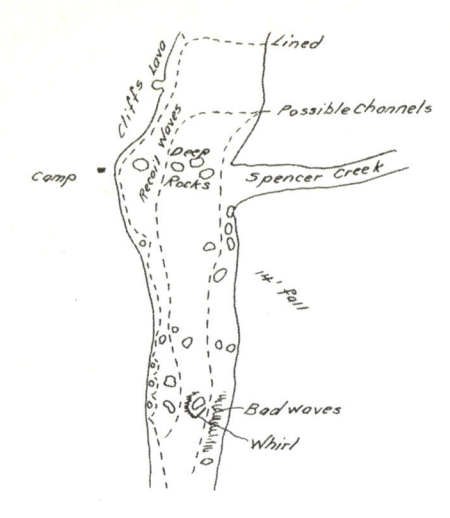

Sketch of Lava Cliff Rapid by Leigh Lint showing where the 1923 USGS expedition lined their boats in addition to the potential runs identified by Blake and Lint.

Lint diary, p. 34, courtesy of George Lint.

is not clear but one infers that he ran the rapid which the writer doubts.[12] Stone lined the boats past the rapids holding the ropes from the cliff above.

BLAKE: We traveled a mile and a half this morning before reaching the rapid known as the Bald [*sic*] Escarpment rapid. The *Glen* was the last boat to land and the first words which reached our ears were "She's sure a bad one, we'll sure have to portage." The next I heard was an opposite opinion from Lint who said, "You and I can run this easy, there's a good channel next to the wall." This opinion I shared as soon as I had studied the details of the course which he had spoken of, but with the exception of Felix, the others seemed to be against us. In fact, Emery had already decided to portage boats and loads and declared it impossible to run, which of course sounded rather queer considering there were no waves but what were much smaller than we encountered time and again.

LINT: We first lined Emery's boat down from a ledge with a 150-foot rope. Dodge had climbed down the 100-foot cliff with the aid of a rope and stationed himself on another ledge. After I had let Emery and his boat down to where Dodge was I rowed back upstream and then crossed over the cliff to help below. Dodge tied Emery's boat and then Emery tied a canteen onto a long string and

---

12. Actually, Stanton did not run Lava Cliff Rapid in March 1890, lining it instead; Smith and Crampton, *The Colorado River Survey*, 248. Unbeknownst to Birdseye, George Flavell also successfully ran this rapid in 1896, which he believed to be "as dangerous as any on the whole river;" Flavell, *Log of the Panthon*, 70. If Birdseye had known these facts, he might have trusted his head boatman less and listened more intently to those two junior boatmen, who saw the run that Kolb refused to see.

let it float down around the point of the cliff. After much trouble we were able to throw a weighted line over Emery's string and pull it in. Emery fastened his end of the string to his painter and Dodge let him down very slowly. When he got to the point of the cliff we pulled his rope in and then pulled the boat in to the small eddy. We unloaded the boats there and carried the load around the ledge about a hundred feet to where we were to camp. We then lined Emery's boat down to a point just below the huge boulder in midstream but not quite to the end of the rapid but past the rocks. We pulled the boat up on the rocks so that it wouldn't pound itself to pieces. From this point we picked out two possible channels through the rapid and Emery decided to let us run it in the morning. By this time it was 4:15 so we ate lunch. Freeman said that it was safer (all agreed) for him to follow with his boat as we had done with Emery's so with slight variations from our former procedure we got his boat down as far as the small eddy when dark overtook us.

BLAKE, REPRISE: Needless to say Lint and I were much relieved to know that we would not have to go thru the tedious maneuver of lining and portaging our boats which being already rather rotten under water, would probably be much worse off after such a procedure.

LINT, REPRISE: The Colonel and Stabler have been working on a dam site all day and La Rue has been taking pictures. Burchard and Moore have been working up Spencer Creek (on left side at head of rapid) so we have been short handed [in doing the portage]. It being too much of a job to carry our beds over the cliff Blake and I decided to sleep up at my boat. With the aid of a flashlight we were able to get back up to my boat without hitting all of the cactus in the way.

BLAKE, REPRISE: ... as it became dark and no sign was seen of the wanderers, Lint and I took the *Boulder* and crossed to the other side of the river, intending to sleep there near the two boats, and as we landed we saw a light at the mouth of the creek, so we knew that the two surveyors had returned. They were tired and hungry so we got out some raisins and pork and beans which, together with a couple of pancakes, which they had left from their lunch, comprised their supper. We did not try to recross the river but made down our beds, Moore using the one belonging to Dodge, the rest of us using our own. We then built a big bonfire of a pile of driftwood and semaphored to the camp opposite that all was well. Several messages were sent and answered, we on the south shore being plainly seen in the glare of the fires, while a flashlight was turned on Emery, who was sending most of the messages. We went to bed in good season, and all night could see the reflection of a light upon the other shore, so we knew that someone was keeping watch against a possible rise in the river, which would endanger the boats. By the same sign we also knew that there was somewhat of a nervous tension in the other camp.

October 12, 1923 — 12,600 ft³/s

STABLER: We left our boats in a rather precarious situation last night if a flood or even a rise of a few feet in the river should take place, so we set a watch all night in two-hour shifts. I was on from ten to midnight. No change in water level all night.

BLAKE: Before crossing the river to breakfast Col. Birdseye held a rod for Burchard, who set up the instrument and took a reading from him. We then crossed the river, climbed up over the deposit of lava between the boat landing and the kitchen which Leigh and I had traversed the night before by the light of a flashlight, and were soon eating mush without milk as we were running short, and bacon and hotcakes. After breakfast the *Grand* was lined and portaged down to a more or less quiet mooring below the *Marble*. Leigh and I were taking a last look at the channel, when Head Boatman Kolb told me that our boats would also have to be portaged as "The colonel refused to permit me to sanction the running of the rapid." This last decision does not sit well with Leigh and I, as we feel that it will damage the rotten bottoms of our boat much more to pull them over the rough rocks on the surface, than to make the run thru. In fact one of the party [probably Dodge] declares that he could swim thru the channel outlined by Lint and myself, and we have no doubt but that he could, but it looks as through the accident of the day before has caused some of the party to be unduly cautious. Wild guesses had been made as to the fall of the rapid, none of them being less than twenty feet and some guessing over twenty-five. So ran the unleashed imagination, while the actual measurement of a fall of only fourteen feet.[13]

---

13. The crew showed that they were unduly cautious, even spooked, by the Colorado River in several ways. They posted watches for a potential rise in the river, remembering their experience at Lava Falls. Birdseye would not allow the two junior boatmen to run the rapid despite acknowledging their obvious skills, in part owing to the previous day's flip and loss of limited equipment and data. Finally, they succumbed to exaggeration of the size and fall through rapids, losing their scientific objectivity despite being in possession of surveying instruments that would answer the question at hand. By the end of the day, they definitely knew one thing: the last of the large rapids was behind them.

Sketch of Leigh Lint
by R. C. Moore.

Edith Kolb diary, Emery Kolb Collection, Box 14, Folder 1756, courtesy of the Cline Library, Northern Arizona University.

BIRDSEYE: Blake and Lint were determined to run the entire rapid and had some disagreement with Kolb over the matter. The writer considers the rapids worse than any in the canyon section except Lava Falls and was forced to issue positive orders that the *Glen* and *Boulder* must be lined and portaged in the same way as was done with the other boats. The two boys were somewhat disgruntled as this was the last bad rapid on the river and they thought they saw a safe way through. Both are absolutely fearless and exceedingly skillful boatmen and no doubt might have taken the boats through safely however, Kolb had found a safe way to line and portage and at the risk of losing one or more boats so near the end of the voyage was too great.

LINT: Lined Freeman's boat down this morning. Lined the *Glen* and the *Boulder* following the same plan as we did yesterday except that with my boat (the last one down). I ran from the point where I lined the others down to where Dodge was without any help from above. The river is very swift along the cliff so we had to be careful. Had all the boats together and the equipment portaged by 12:30. Started on down the river about 2:30 and made 1.8 miles. Camped on the right hand side of the river.

BLAKE, AUTOBIOGRAPHY:[14] Both Freeman and Emery had agreed that they thought Lint and I could make a successful run, but that

---
14. Blake, "As I Remember," 171.

The portage of reportedly 2 tons of equipment around Lava Cliff Rapid, October 12.

E. C. La Rue 445, courtesy of the U.S. Geological Survey Photograph Library.

they thought it best that they should not try it. Now we were denied the chance. Naturally, the event would be recorded in any account published. It would undoubtedly seem odd that the two low paid boatmen could make the run while the head boatman and a higher paid boatman did not try it. It was here that the friendship of Lint, myself and Kolb was dampened. Kolb had usually slept beside Lint and I with a common tarpaulin over our beds. Now he felt our resentment over not having been allowed to run the rapids, and thereafter slept apart.[15]

October 13, 1923 — 12,600 ft³/s

LINT: Rapid No. 62 [Surprise Rapid] is about 2500 feet below our last night's camp. It has three boulders in it on the right side at the head and one opposite in midstream. There is also one rock

---

15. The two junior boatmen, who were firm allies of Emery Kolb to this point in the expedition, lost faith in their head boatman. As Birdseye's narration clearly indicates, the skill level of the two junior boatmen, as well as Dodge, far exceeded that of either Kolb or Freeman.

about fifty feet from the right bank and below the first ones. Ran it by going to the left of the first rock, then to the right of the next one, and just missing the lower one by going to the left of it. There is a clear channel down the left side but it is full of large waves and we didn't want to get wet. 8 feet fall. The Colonel worked up the side canyon on the right side at the head of the rapid. Rapid No. 63 [Lost Creek Rapid] with 8 feet fall is about a mile below No. 62. It has one rock on the left side just below the head and one deep submerged rock almost opposite near the right bank. Ran it by going between the two rocks and then pulling to the right to avoid the largest waves.

BIRDSEYE: ... the party passed several interesting side canyons one of which enters from the right 24 miles below Diamond Creek and was named Bottle Neck Canyon [Surprise Canyon] on account of the exceedingly narrow entrance into the Grand Canyon and the deep wide valley above. Fresh cattle tracks were found within a quarter of a mile of the river and a small amount of clear running water flows into the river.

FREEMAN: Overtaking the rod and instrument, we landed on the bar where a canyon and small stream came in from the left. I rodded for Birdseye while he carried a line to the 2100-foot level. Burchard, saying that things began to take on a familiar aspect, pushed off into the riffle below. An hour later we found him with a grin on his face, pointing to a monument across the river as one which he had erected at the highest point reached by his survey of 1920. That ended the work of carrying the line down the river. Just above Burchard's monument we came through one of the queerest riffles on the river, or at least it is such at this water. A boulder bar on the left throws the main current at right angles across to the right wall, where it conflicts with an opposing set of currents in a way to cause a heavy reverse wave like that below a large submerged rock. In trying to work to the right of this apparent obstruction, the boat literally stopped and held like a hovering bird. At the end of five or six seconds of indecision, she yielded to the persuasion of the oars and dropped on down over what turned out to be no more than sloppy water.

BIRDSEYE, REPRISE: Burchard ... checked his elevations with an error of only 4½ ft, a remarkable feat when one considers that differences of elevations were often determined by vertical angles with sights as long as one half mile. All were pleased with the results and glad to have the actual survey work over with nothing ahead in the way of bad rapids and only a 200 mile pull to the Needles.

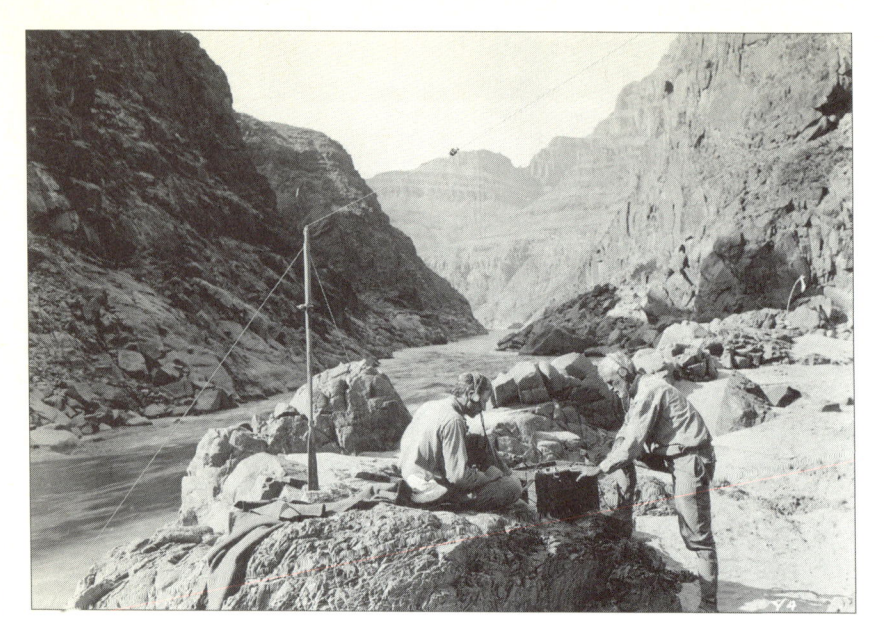

Herman Stabler and Claude Birdseye with the radio setup at Devils Slide Rapid near the western end of Grand Canyon.

Emery Kolb Collection, NAU.PH.568-5239, courtesy of the Cline Library, Northern Arizona University.

STABLER: Today we ran as far as Devil's Slide dam site and camped for the night. Tomorrow we will make the survey. It is one picked from Burchard's map and specially mentioned by him. It has an exceptionally good spillway from the canyon of a tributary creek, and while it may not fit in with the ultimate scheme of development there is a chance that it may.[16]

FREEMAN, REPRISE: Our camp is on a broad, low bar, covering several areas, and which forms the right hand channel around a granite island at high water. Devil's Slide Rapid, just below, is a fairly rough piece of water, and Burchard's feat in bringing one of his open boats up over it was a very creditable piece of work. The radio is set up on top of the granite island, with an aerial to the

16. At this point, the survey work of the 1923 USGS expedition ended, although La Rue would still make some official photographs. They had gone 255 river miles from Lee's Ferry and passed through all the significant rapids of Grand Canyon, the end of which was another twenty-two miles downstream. They shifted mode to what is now known as "a deadhead," or boating to their takeout as quickly as possible. They had chosen to take out at Needles instead of closer places simply because of the presence of the railroad terminal, which would facilitate shipment of heavy gear away from the Colorado River.

righthand cliff. Stations in Los Angeles, Salt Lake and Chicago have been picked. We had KFI of Los Angeles for the first time. As usual, KHJ was the strongest. We learned that the Yankees evened the World's Series by winning today; also that there is a severe brush fire in Eagle Rock Valley.

October 14, 1923 — 12,500 ft³/s

STABLER: At work on the dam-site survey all day. Elevation of saddle 1258, 1242 on river side—700 feet wide. Section 590 feet at 1350. W.L. 1037. Grub is getting short. We are down to a very restricted diet. A little bacon left that will go tomorrow leaving us without meat. Plenty of flour and sugar and oatmeal and a little bit of a few other things. We hope to shoot a few ducks. Then we may be able to get a little stuff from a ranch or two near Virgin River some 70 miles below. We can get through on what we have if necessary but naturally would prefer more variety. Will make a big day's run tomorrow.

LINT: All of us except Freeman put an oar lock for a steering oar on the stern of our boat and Emery put a skag keel on his boat [for better tracking in the flat water downstream]. I painted the names of all the party on a low wall above camp about 75 feet above the water. For the last two days the canyon has been full of smoke, coming from some forest fire no doubt.

October 15, 1923 — 12,600 ft³/s

STABLER: We got an early start and ran to within 1½ miles of Pierce [Pearce] Ferry,[17] abandoned and desolate, the mouth of the Grand Canyon, by noon. This was about 20 miles for a morning's run. I rowed about half the time, taking turn about with Emery. On our way down Emery shot two ducks and I shot at some without fatal results. The break from the Grand Canyon into open country is very sudden and striking. We are again in a semi-canyon region though the canyon walls are only a few hundred feet high and are broken into hills and valleys in many places. We should reach Virgin River, possibly Boulder Canyon, tomorrow.

LINT: In the second rapid about a mile below camp Burchard was trying to hold the steering oar out of the water and when the boat hit a whirl he fell off in the river. He grabbed the life line and climbed back on the boat. His hat fell off and we couldn't locate it again. Freeman fell into the rear the first thing this morning and

---

17. The 1923 USGS expedition may have been the source of misspelling of Pearce Ferry, named for Harrison Pearce, who founded the ferry in 1876; Brian, *River to Rim*, 134.

The ruins of Pearce Ferry, now submerged beneath Lake Mead. La Rue photographed this view on October 15, 1923.

E. C. La Rue 771, courtesy of the U.S. Geological Survey Photograph Library.

---

never caught up with us until after lunch. The Colonel spent about two hours and a half on a poor dam site just below where we ate lunch. Emery went ahead and killed 17 quail and a coyote; saw two other coyotes. Made 29 miles today and camped at the mouth of Grand Wash.

FREEMAN: There are several bad rocks in Devil's Slide Rapid, but a broad channel through. Not far below here is another riffle—a swift mid-stream chute with several side-winding waves at the foot of the V. Burchard christened this Triumphal Arch Rapid, so it was not inappropriate that it should be he who was thrown off the stern of the *Boulder* in running through. He was back again in a moment, but thoroughly soaked. The *Grand* skimmed the tops off of several waves in her passage, and both La Rue and Moore were slightly moistened. Travertine formations, with running springs, were frequent, and over one on the right bank a fine waterfall tumbles. This gave the name to the rapid below—all nomenclature Burchard's. Hell Diver Rapid proved a series of long winding riffles, with a total fall of 20 feet. The head was fairly rough, with a somewhat restricted channel to the left. Just below this rapid— or possibly included as a part of it—is a boulder riffle setting with great force against the right wall. I crowded the bar to the left as closely as possible in heading in, but even so the *Grand* led a merry dance down the side of the back-thrown wave. In this, or a similar rapid below, Stabler says that Kolb had rather a messy time of it, shipping half a cockpitful of water. After the granite ran out the fall of the river came mostly in long swinging boulder riffles, easy to run but with enough current to make progress rapid. There was one riffle, like that above Burchard's upper monument, where the river turned on its back and ran under itself like a writhing snake, giving the effect of a rocky obstruction which really did not exist. This one, like the other, had no teeth in it, though it looked too bad to take liberties with. The "Grand Canyon" character of the gorge held to the very last. Seldom have we seen finer walls than those rearing above us even as a doubled bend showed a greatly lowered horizon ahead and a hill with a rounded top. The river breaks out from under the great Colorado Plateau two miles above the Old Pierce [Pearce] Ferry, and seven or eight from the mouth of the Grand Wash. As we ran out from under the great wall of the Grand Wash Cliffs, the granite came up again for a short distance. A mile below this final narrow section we pulled in for a look at what remains of Pierce [Pearce] Ferry, the farthest-up crossing of the Lower Colorado. Only traces of the former cables, roads and the walls of a building of gypsum blocks survive. The Yankees won the series by taking their fourth game today, radio announces.[18]

BLAKE: Dodge and I took turns rowing and were soon far ahead of the others. A number of small rapids and riffles were run and finally, thinking that some of the boats might have had trouble, we stopped and Dodge made himself some coffee while we waited for the other boats. The *Marble* and *Boulder* were seen in the distance after a half hour's wait and we fell in behind them as they passed. While waiting for the surveyors I rigged up a sail for my boat and sailed up the river in the still water. We caught up with the other boats about 5 o'clock. Supper was late but was well worth waiting for. Roast quail, duck soup, peas, curry gravy and hot biscuits.

October 16, 1923 — 12,500 ft³/s

FREEMAN: As La Rue is continuing his photographic work to the mouth of the Virgin, the *Grand* pushed off early this morning. The succession of riffles round the wide bar of the Grand Wash have

---

18. This World Series, an issue of interest to the crew of the 1923 USGS expedition, was played at the pinnacle of Babe Ruth's career. Ruth led the Yankees to victory over the Giants. "Slugging Yankees Wallop Giants," *Los Angeles Times*, October 12, 1923 (accessed June 13, 2006).

some good sized waves at the head, but we ran through without splashing into the open forward hatch, where Moore now squats with the laudable dual intention of increasing comfort and decreasing wind-resistance. We landed below to make a final picture of the Grand Wash Cliffs, with the break of the gorge of the Grand Canyon still marked by a dark slash of shadow near the rim. Iceberg Canyon is well named, the gray-green pinnacled formations strikingly suggesting those of a great floating ice island. The current continued favorable, and Moore and La Rue—pulling double—made good time in their first turn at the oars. About ten, after passing a deserted mine and mill on the left bank, we came to Grigg's [William Grigg] Ferry, to find a group gathered around our two leading boats. All too late we discovered skirts flapping in the wind, and as a consequence committed the unpardonable faux pas (in this region) of landing without our shirts. To make matters worse, both La Rue and Moore had considerable gaps in the seats of their pants. These latter they managed to keep out of the picture by advancing and backing as in the presence of royalty; but the shirts were irreplaceable, once we had landed through the mud. The ladies were terribly embarrassed; likewise the men, one of whom would not even look up when introduced. The scandalized humans proved to be the Smiths, who run the ferry. They have been in the country 28 years. One of the girls was a bride; the other her younger sister. Our deshabille was a distinct source of uneasiness to them as long as we stayed, and the worst horrified of the men did not even offer his hand at parting. Darned funny!

BLAKE: We left camp at 7 o'clock this morning and reached the Smith ranch at 9:30 where we bought some canned goods, potatoes and dry beans. They gave us some home grown dates and told us that they raise an abundance of figs also. After leaving the ranch we pulled down the river several miles, before stopping for lunch running Hualapai (Walapai) Rapid which is supposed to be a bad rapid, having taken toll of several lives from the local residents. We did not even remark that it was a rapid, and not merely a big riffle until some one told us that it was the notorious Hualapai rapid.

FREEMAN, REPRISE: Running through a broadening valley in a widening river, we came to Temple Butte about three o'clock. Rising from the right side of the river to a height of perhaps 300 feet, and almost four-square as seen from above, this sand-and-conglomerate formation is one of the most striking landmarks on the Colorado. To me it was strongly suggestive of one of the great sun-dried brick ruins of ancient Babylonia and Assyria. La Rue and I landed and took several pictures. We passed Temple Bar two miles below the butte. Considerable hydraulic operations have been carried on here at a comparatively recent date, as the machinery and other installations look fairly modern. Napoleon's Tomb, a mile farther down, is a far less prominent monument than the Temple Butte, and the name is rather far fetched. We were in sight of the mouth of the Virgin before sundown, landing to camp on the beach at the base of a line of cliffs on the right. At a point very near here Powell's pioneering voyage was finished in September 1869. The only evidence of former life at this historic point is the ruin of a stone house just above the mouth of the Virgin. It was probably built by the Bonelli's, ferry and cattle people.

LINT: Passed the old Harry Armitage place about a mile and a half above the mouth of the Virgin River. Made 38.2 miles today and camped about 1,000 feet above the mouth of the Virgin River. Emery killed 5 ducks and 3 quail today. Between the whole outfit we saw seven coyotes today.

STABLER: After the first 11 miles tomorrow I will be on ground passed over last year. About 11 o'clock we reached Smith's Ranch at Grigg's Ferry and saw the first people (except our supply men) since leaving Bass Trail. The recent flood on the river washed out the ferry and they had only just finished a small boat to replace it temporarily and had no oars, but had made some makeshift ones out of planks.

October 17, 1923 — 12,400 ft^3/s

STABLER: Passed Boulder Canyon early in the day. The old USRS camp was deserted, except for a stream gager, Jackson, and his assistant. The assistant was making a measurement as we passed under the cable and told us the discharge is about 11,000 second-feet, a fine stage to travel by. The recent flood, he told us, was 125,000 second-feet, which seems too high for Boulder Canyon. Below Old Colville our boat stopped at the camp of the Stetson outfit. Nobody but the caretaker was at home. He invited us to dinner and on our refusal invited us to have a drink of real cow's milk, which we did. They have quite a nice layout, including a dog, 4 cats, a cow, and two calves. We found the Black Canyon camp deserted also. Below there we passed the old deserted mining camp at Jumbo

Boulder Canyon Dam site, located in Black Canyon and the present site of Hoover Dam, photographed in 1922 by the U.S. Bureau of Reclamation.

Courtesy of the U.S. Bureau of Reclamation.

Wash and finally camped about 2 miles above the old El Dorado Ferry. We now have a pretty good chance to reach Needles the day after tomorrow.

BLAKE: Dodge and I went ahead this morning taking the short gun with us. We only got one duck before noon and two after noon. We stopped at the head of Boulder canyon where some surveyors were at work. We had heard of two big rapids in the canyon but found only a couple of riffles. I put on a big dutch oven full of beans to cook all night,[19] as we ate nearly all those we cooked yesterday.

LINT: At about 11:15 we passed the ruins of old Fort Colville which was built by the Army in 1840. The old rock houses and fences are still standing. About a half hour later we passed a camp of some engineer and he had a radio up and a flat bottom boat that was propelled by an aeroplane motor and propeller. Passed an

---

19. In his autobiography, Blake gave his bean recipe, which included strips of bacon and a bottle of ketchup. "That mess of beans got me a job as cook on a later government trip, when Stabler told about the delicious beans I could cook. He never knew whether I could cook anything else, or not." Blake, "As I Remember," 175–76.

old Reclamation Service camp at the head of Black Canyon. This canyon is 28 miles long and we camped at the mouth of it. Except in the two canyons the river has been broad and the current slow. Made 51 miles today, a big day's run.

October 18, 1923 — 12,300 ft³/s

BLAKE: At 7:30 we were all ready to start. We had rowed a mile or so when a breeze again sprang up and sails were spread on all but the *Glen* as I had determined not to waste time on it again, but as the wind increased Dodge said, "I am game if you are," so we again put up our sail using an oar for a mast, and a four by seven bed tarp for a sail. As the wind kept steadily blowing southward we rigged up a topsail by sticking another oar about four feet higher than the first. We soon were making good time, and tho the other boats were adding more canvas we drew away from them until we were out of sight. Twice we drifted into difficulties in the way of a sand bar and some shore bushes, but were soon on our way again. We passed Tristate, Nev., about 2 o'clock. Our only trouble came when we struck a long wide bend in the river where it ran directly west. Here the wind blew at right angles to our course and drifted us to shore, so we had to haul down our main sail and row to deep water. A half mile below this big bend we halted and waited for the other boats as we had two beds which did not belong to us or we should have been able to reach Needles. During our two hour wait we made some beef tea and ate some raisins, as we had not waited to lunch with the other boats. We made camp on the west bank of the river where a big drift covered sandbar extended a hundred yards from the mosquito infested willows.

FREEMAN: With the favoring wind holding and promising to strengthen, canvas was broken out on the whole fleet as soon as we pushed off. Moore manipulated his bed cover on the triangle of the folded rod with good results all morning—rather a Malayan felucca effect when well filled. At the noon lunch halt we rigged an oar for a mast and hoisted upon it a huge square-sail improvised from La Rue's bed canvas. It had a powerful drive in the heavier gusts, keeping Moore busy trimming and La Rue steering. Except for an occasional dig with the oars to help steering, I had nothing to do but sing out orders as to the course ahead and to pull out of some hole now and then where there was no steerage way.

Packing the boats (from left to right, the *Marble*, *Grand* and *Glen*) and equipment at Needles, California, the terminus of the 1923 USGS expedition. The man at the far left is probably Frank Dodge; immediately to his left is probably Felix Kominsky.

Emery Kolb Collection, NAU.PH.568-5215, courtesy of the Cline Library, Northern Arizona University.

October 19, 1923 — 12,100 ft^3/s

BLAKE, AUTOBIOGRAPHY:[20] I again took the lead, rowing hard to keep well ahead of Freeman, for we young bucks were determined he could not brag in his account of the last days of the trip that he had outstripped the others.

LINT: Pulled into Needles at 9:30 this morning and had the boats and all the equipment hauled up to the freight depot by 12 o'clock. I helped the Colonel and Burchard pack the instruments and some of their equipment this afternoon.

BLAKE, REPRISE: It took most of the afternoon for the party to shave and make themselves presentable. Although there was a great change in all of us, from the river garb and unshaven faces, the greatest change was in our good natured cook, Felix. When he reached town he wore a bushy, red beard. His straw hat which had not increased in beauty by its frequent duckings in

---

20. Ibid.,178.

the muddy Colorado, covered only a portion of his long colorless hair. A blue denim jumper sufficed as a coat, and his checkered breeches were very greasy from much wiping of the butcher knife after cutting bacon and ham. He disappeared for about two hours but was not taking a beauty nap during the time as was evidenced by the startling change noted as he strolled into the lobby of the El Garces hotel. Freshly shaven, with a long cigar tilted skyward from one corner of his mouth, he strolled into our presence shod in latest oxfords. His rotund figure had been fitted by a stylish coat sweater and neat trousers. He looked, as some one said, "like a million dollars," all but his hat, which was a fuzzy skyscraper six gallon type, worn by movie cowboys, but despite its incongruity with his other raiment, it seemed just what Felix should wear. We who know of the average camp cook, decided to show our appreciation of the never failing good humor of our exceptional friend and so made a pool of substantial size and purchased a diamond stick pin, which was presented to him at a special dinner in the evening.

FREEMAN: Considerable changes in the river were evident since last year, the most notable of these being at Needles. Here we found a bar 200 yards wide shutting off the bank against which we had landed last season. This, we are told, occurred during the late flood, which also carried away most of the boats owned in this vicinity. The sand bar at which we landed proved hard if damp, and it was possible to drive a truck to the water's edge. Before one o'clock all of the boats were on the station platform and their loads at the hotel or in a warehouse. Tomorrow is to be spent in packing and shipping outfit and boats and cleaning up generally. There will be a general exodus tonight and tomorrow. Roger Birdseye arrives on an early morning train to help the Colonel clean-up business affairs. No one in the party itself has had more to do with its success than the "Manager of Land Operations."

BLAKE, REPRISE:[21] The first night in town only Dodge and I went to the movies. As we were walking down the darkened aisle, Dodge suddenly stopped. "Blake," he said, pointing to the screen, "That's us." Sure enough, there we were, plunging down the midst of Hermit Creek Rapid. It was a Fox Movietone, being shown before the main picture.

---

21. Ibid.

October 20, 1923 — 11,800 ft³/s

LINT: Packed up the equipment and shipped it; also loaded the boats. Emery and Burchard left for their homes last night and Felix left this morning. Stabler, La Rue, Moore, and Freeman pulled out this evening.

BLAKE:[22] ... we were entertained by the Chamber of Commerce at a banquet. One of the men at table was the editor who had written such a vivid description of the "Lost Survey Party" and how parts of their boats had been seen floating down the Colorado during the big flood. Dodge asked him about the story and ridiculed him to his face, to the embarrassment of the Colonel.

October 21, 1923 — 11,500 ft³/s

LINT: The rest of us, with the exception of Mr. Birdseye, left this morning, thus completing the breakup of the "Gang."

KOLB, LETTER:[23] Our trip ended Oct. 19. I cannot say but what I was glad to get out, as the slowness gets one's goat as the surveyors never made ten miles a day, usually 5-4-3-2-1 and at side canyons staying at the same camp two nights. However it was as a whole a great trip, though with such a big party, ten at first, and eleven from Bright Angel, it made the work heavy for the ones who did the work. Referring to your surprise at an army officer tolerating Freeman on the trip, I will say you are not alone in your surprise. Not so much the army officer, for Col. Birdseye was not an old veteran, but at he or anyone tolerating him. My only explanation, which is entirely presumption, is the fact that in 1922 Freeman was with the Colonel on the short trip from Hall's Crossing to Lee's Ferry; and desiring to get on the 1923 trip, would make the best showing he could on that short trip, which he apparently did. More than that, Freeman wrote up the trip in monthly installments for the *Sunset Magazine*, and while I have a great deal of respect for the Colonel, I could see that it pleased him very much to have anything [about] the U.S.G.S. in print, regardless of the fact that Freeman openly admitted to us that many of his statements in the articles were damn lies; stating that he never stuck to the truth when he wanted his stories to sound funny. Mr. Stabler and Mr. La Rue were also with the party, though conservative in their expressions, stated that there was much in the articles that never happened. Knowing these facts before

---

22. Ibid.
23. Emery Kolb to unknown, probably Julius Stone, November 4, 1923, box 111, folder 16, Marston Collection.

leaving Lee's Ferry and then finding out what a lazy foreflusher he [Freeman] really was, caused a feeling with the whole party which I presume compared with the feeling you had for Cogswell.[24] After starting on the trip, he could not carry loads on account of a shoulder which had slipped out of place nine different times. He could not rod in emergencies where there was any climbing or long walking on account of his two "busted ankles" injuries from football days. Yet he would forget about his shoulder and show how (fast?) far he could go through large boulders, and near the end of the trip was betting on his jumping ability. His ego, laziness, and selfishness could only be compared to Griggs or possibly La Rue, the hydraulic engineer who rode with Freeman, both of which tried to impress the rest of the party that the other should not have been on the trip. Colonel was a good chief. The hardest worker I have seen for many a day and about the only complaint any of the party had was when they were made to do Freeman's work just because he was too big slow or thought too slow. After working all day I have seen the Colonel mop out Freeman's boat while Freeman set on the bank watching him do it. Some brotherly love!

---

24. Julius Stone selected Raymond C. Cogswell, his brother-in-law, to serve as photographer on his 1909 Green River–Colorado River expedition. Stone did not care for Cogswell before the trip, and disliked him even more afterward. Stone, *Canyon Country*.

8

Aftermath

Politics and the Strident Hydraulic Engineer

Upon returning to their homes, the members of the 1923 USGS expedition to Grand Canyon were drenched in a maelstrom of hyperbole. Just as Blake and Dodge felt when they saw themselves on the movie screen in Needles, everyone must have found the attention both astonishing and amusing. Dozens of newspapers and magazines across the country carried stories about the expedition, laden with adjectives that spoke of the purported hardship and danger, real and imagined, that they had faced. The *Washington Post* ran a photograph of Birdseye, looking steely jawed and intense, with the caption, "C.H. Birdseye, who, with his party, was believed lost in the Colorado River, returned to this city yesterday."[1] The *Los Angeles Times* interviewed La Rue, who obligingly described the flood scene but also managed to note that the expedition had been successful in its purpose: "We were sent to secure certain information for the government, and we got it."[2]

A press release based on Birdseye's story was distributed in November, and another round of sensationalistic newspaper headlines followed. "Daring Scientists Conquer Grand Canyon's Wild Rapids: Hurled in Frail Boats Down 280 Miles of Savage Stream!"[3] and "Chief of Canyon Explorers Once Reported Killed

---

1. Photo, *Washington Post*, November 3, 1923 (accessed March 8, 2006). See photo on page 59 above.
2. "Canyon Flood Peril Pictured," *Los Angeles Times*, October 22, 1923 (accessed March 7, 2006).
3. Stuart M. Emery, "Daring Scientists Conquer Grand Canyon's Wild Rapids," *New York Times*, November 11, 1923 (accessed March 8, 2006).

La Rue, Birdseye, and Stabler standing near the *Grand* in front of the Interior Department, Washington D.C. January 12, 1924.

Underwood & Underwood photograph, 36177RU, courtesy of the Library of Congress.

Tells of Perils: Risked Lives for 450 Miles to Make Maps!"[4] were two examples. The more subdued professional journal *Engineering News-Record* entitled its report, "Birdseye Party Completes Survey of Grand Canyon,"[5] while the *Mining Congress Journal* chose to emphasize something different entirely with its headline: "Grand Canyon Survey Discloses Vast Power Resources."[6] The *Los Angeles Times* and KHJ radio were pleased that "KHJ came to be accepted as a regular functioning unit of the outfit,"[7] thereby firmly establishing the reach of commercial radio into the wilderness of the United States.

4. "Chief of Canyon Explorers Once Reported Killed Tells of Perils," *Washington Post*, November 11, 1923 (accessed March 8, 2006).
5. "Birdseye Party Completes Survey of Grand Canyon," *Engineering News-Record* 91 (November 15, 1923): 808–11.
6. "Grand Canyon Survey Discloses Vast Power Resources," *Mining Congress Journal* 9 (December 1923): 460–61.
7. "Radio Passes All Barriers," *Los Angeles Times*, August 4, 1924 (accessed March 8, 2006). This article was in response to Freeman's *National Geographic* story about the expedition, which praised KHJ.

Much like the aftermath of modern-day space travel or sports exploits, mementos of the expedition were sought for preservation by museums. The *Glen*, *Marble*, and *Boulder*, which were owned by Southern California Edison Company, were shipped back to Los Angeles. The *Grand*, the only boat outright owned by USGS and rowed by Lewis Freeman, was donated to the Smithsonian Institution, where it remains.[8] The *Glen* has recently been renovated as part of the Grand Canyon National Park Foundation's "If Boats Could Talk Conservation Project" and is at the South Rim of the Canyon.[9] The *Marble*, as of 1983, was on display at the Southern California Edison Big Creek Powerhouse No. 1 near Big Creek, California.[10] The whereabouts of the *Boulder* are unknown.

Hidden from general knowledge was the controversy that continued to swirl among the crew members. Birdseye would try to control it, but discontent among his former crew would fester for many years; in some cases, animosities would last a lifetime. Careers were affected, positively or negatively. Freeman, the butt of so many jokes and malicious comments during the expedition, would justify his inclusion on the river trip many times over. Kolb, bitterly upset at the spread of images of Grand Canyon that he could not profit from, attempted to make his version of reality a congressional issue. La Rue, pushing his agenda for water development, ultimately would forfeit his career. The steady Burchard, the venerable Birdseye, and Moore would retire to the sidelines, but they would earn accolades, and deservedly so, from their agency and peers.

After they stepped onto the train at Needles, the crew members never reassembled in one place. Some were lost to history, returning from an extraordinary moment in the national spotlight to ordinary lives. Small groups continued to work together, producing trip reports, maps, and monographs that detailed their scientific results as well as the near-heroic aspects of the expedition. Their exploits would appear in *National Geographic*, newspapers, several journals, and yet another book by the prolific Freeman. Birdseye's college fraternity gushed over his leadership of the expedition: "This dangerous and nearly unknown stretch of water has for years baffled the explorer, the scientist, and the map-maker, and it has remained for

---

8. The *Grand* is in the collections storage of the Smithsonian's National Museum of American History and is unavailable for viewing.
9. "If Boats Could Talk Conservation Project," http://www.gcnpf.org/projects/historicboats.html (accessed September 13, 2006).
10. Myers, *Iron Men and Copper Wires*, 180.

a Kappa Sigma to head one of the most hazardous expeditions sent out by the government in recent years ..."[11] The trip even served as the inspiration for a work of youth fiction, *Colorado River Boy Boatmen*, which included figures modeled after Birdseye, Kolb, and Dodge.[12] In addition to a monument on the South Rim of Grand Canyon, some very large monuments to the 1923 USGS expedition now impound reservoirs on the Colorado River, and perhaps of more importance, the Colorado River freely flows past most of their proposed dam sites.

The Crew Disperses

Claude Birdseye

Birdseye showed the highest levels of leadership by holding his fractious expedition together to its successful conclusion. He faced new challenges: he needed to justify the expensive trip and get the scientific data compiled and published; and second, he needed to organize the publicity to feed a national audience eager to learn of the experiences of the expedition. His most pressing need may have been to deal with the ongoing interpersonal conflicts, which threatened to go public. Birdseye took to the typewriter, composing articles for *The Military Engineer* and *The Geographical Review* (with Moore).[13] In addition to the much-quoted press release, Birdseye gave a radio address that centered on the usefulness of the radio to the expedition.[14]

To Birdseye, the 1923 USGS expedition was merely another achievement in a long and illustrious career. In 1924, the American Geographical Society presented Birdseye with the Charles P. Daly Medal for his distinctive scientific record, of which the 1923 trip was a large part. He was awarded an honorary doctorate by Oberlin College, his alma mater, eight years later.[15] In 1928, he submitted

---

11. Martin, "Conquers Grand Canyon," 322.
12. John F. Cowan, *Colorado River Boy Boatmen* (San Diego: Stockton Press, 1932). The book includes a 16-year-old named Larry "Carrot-Top" Dennison, who, with a friend, runs a photographic studio at Grand Canyon. Larry is invited to join a Grand Canyon River survey party by Keola Akoka, who, like Frank Dodge, is a strong swimmer and, unlike Dodge, speaks pidgin English. Akoka previously worked in Hawaii with "Colonel Buckeye," the leader of the Grand Canyon trip.
13. Birdseye, "Surveying the Colorado Grand Canyon"; Birdseye and Moore, "Boat Voyage Through Grand Canyon."
14. Birdseye, "Radio in the Grand Canyon."
15. Wilson, "Memoir of Claude Hale Birdseye," 1551.

a letter of resignation to USGS Director George Smith: "I have the opportunity of becoming the administrative and technical head of a new aerial photographic surveying company, which will permit me to give all of my time to the subject in which I am most interested. I have hesitated to accept only because of my keen regret at leaving the organization in which I have served for twenty-five years, most of this period being under the present Director who has honored me with his confidence and given me every opportunity of free and independent technical action. I would now refuse the new position if I were not sure that others in the Geological Survey can perform my present duties with even better results than I have been able to accomplish."[16]

Smith reluctantly accepted Birdseye's resignation, noting Birdseye's typical modesty, and continuing: "Nearly ten years ago, in recommending Colonel Birdseye for this position, I said, 'I believe he is the one man for this place.' The record of ten years has demonstrated the high quality of his service and makes replacement difficult."[17] One of Birdseye's tasks as president of the new company (the Aerotopograph Corporation of America) was the photogrammetric survey of the Boulder Dam site in Black Canyon, which found him in the rugged country of northern Arizona once more.[18]

In joining the private sector and promoting photogrammetry, Birdseye recognized that the technique USGS used in Grand Canyon was obsolete, to be replaced by the evolving science of remote sensing. Aerial photography was in its infancy in the early 1930s; the technique of photogrammetry, or the extraction of quantitative topographic information from aerial photographs, would dominate large-scale topographic mapping for most of the twentieth century. The Depression likely forced a retreat from private industry, and Birdseye returned to USGS in 1932 in a position that was created specifically for him.[19] Initially, he was an assistant to the director, and then he became the chief of the Division of Engraving and Printing, which created the topographic maps that were one of

---

16. Claude H. Birdseye to the secretary of the interior, August 26, 1929, Birdseye Personnel File.
17. George Otis Smith to Ray Lyman Wilbur, September 9, 1919, Birdseye Personnel File.
18. Birdseye published his findings on the Boulder Dam site in 1931; Claude H. Birdseye, "Photographic Surveys of Hoover Dam Site," *Civil Engineering* 1 (April 1931): 619–24.
19. That a position was created for Birdseye during the depths of the Depression is a testament to Birdseye's reputation. For more information on Birdseye, see Wilson, "Memoir of Claude Hale Birdseye,"1549–1553; R. H. Sargent, "Colonel Claude Hale Birdseye," *Annals of the Association of American Geographers* 32 (1942): 309–15; Heinz Gruner, "Colonel Claude H. Birdseye," *Photogrammetric Engineering* 38 (1972): 865–75.

the pillars of USGS. In 1938, he was commissioned as a full colonel—he had previously held the rank of lieutenant colonel—in the Army Reserve's Engineer Corps. The promotion was "in recognition of his increasing prominence in a field of engineering that is so important in time of war."[20]

When Birdseye died in 1941, he was eulogized in both professional journals and in newspapers, many of which specifically noted the 1923 expedition. He was buried in Arlington National Cemetery with full military honors.[21] Birdseye's work was lauded posthumously. In 1975, the American Society of Photogrammetry placed a memorial plaque to Birdseye and the 1923 expedition at Hopi Point on the South Rim of the Grand Canyon.[22] In 1999, he was named one of the "Engineers of the Millennium" by the American Society of Civil Engineers,[23] a high honor, indeed, for the leader of one of the most ambitious surveying projects of all time.

The Other Topographers

Returning to Texas, Roland Burchard joined his wife and three-month-old son, whom he had not seen since the infant's birth in July. The couple had a second child, a daughter, two years later. Burchard achieved a measure of singularity not accorded to his colleagues on the 1923 expedition: he bears a striking resemblance—from his angled features to his leather belt pouch, rumpled field clothes, and lace-up boots—to the surveyor portrayed on the cover of the October 1925 issue of *Scientific American*, shown crossing a seething torrent suspended from a cable, stadia rod in hand, with the fantastically rendered walls of the Grand Canyon looming around him.[24] The painting's caption reads, "All in the day's work." (See the illustration facing this book's title page.) It was fitting: of all the participants in the 1923 Grand Canyon expedition, Burchard, with his Last Chance Rapids cairn in his crosshairs, was probably the most focused on completing the job at hand.

---

20. Wilson, "Memoir of Claude Hale Birdseye," 1552.
21. Obituary, *Washington Post*, June 1, 1941 (accessed March 3, 2006).
22. "Birdseye Memorial Dedicated," *Topo West* (November–December 1975): 11–14. Emery Kolb attended the dedication, as did Karl Sweigart, one of Birdseye's grandsons, and Leigh Lint's son George.
23. Cory Sekine-Pettite, "Engineers of the Millennium," Graduating Engineers Online, http://www.graduatingengineer.com/articles/feature/12-20-99.html (accessed September 19, 2006).
24. Cover, *Scientific American* 133 (October 1925).

Burchard spent the next thirty-two years working for the USGS. From 1930 through 1937, he helped map Los Angeles County, then he moved his family north to Sacramento where he continued to work as a topographic engineer, working in the field as much as possible. From 1952 until his retirement, he was in charge of the Map Distribution Office.[25] Burchard maintained a lasting friendship with Emery Kolb, exchanging periodic correspondence. The two men admired each other; Burchard expressed his fondness for Kolb in letters home, while he impressed Kolb as a "hard, hard worker."[26] Burchard died in Sacramento in 1968.[27]

Unlike La Rue, Burchard quietly supported USGS throughout his life, serving as the ultimate "company man." While Birdseye deserves credit for holding the expedition together, Burchard collected the survey data, performed the quality control, and ultimately linked all the surveys of the Colorado River together. He was one of three men singled out for recognition in the official 1923 USGS press release about the expedition (the other two being Birdseye and Kolb): "The work of the expedition was made successful largely by ... the persistence of the indefatigable Burchard ..."[28] Upon his retirement in 1956, Burchard received a Citation for Commendable Service, which specifically mentioned his selection of the Hoover Dam site and participation in the 1923 expedition.[29] His agency rewarded him for his objectivity with respect to the location and characterization of dam sites on the Colorado River. He privately favored the Boulder Canyon dam site over the high dam in Glen Canyon, later revealing that he argued with La Rue about the subject.[30]

Back in Washington, Herman Stabler resumed his post as head of the Land Classification Board. Colorado River Basin issues continued to occupy part of his time, as he helped oversee completion of various studies and their concluding reports, including those by La Rue. In 1925, he became chief of the Conservation Branch of the USGS. In this capacity, he was responsible not only for land classification, but he also supervised mineral leases on public and tribal

---

25. Burchard, "Life Sketch," 7.
26. Otis Marston, interview with Emery Kolb, Grand Canyon, January 13, 1953, box 111, folder 16, Marston Collection.
27. Social Security Death Index, RootsWeb.com, http://ssdi.rootsweb.com/cgi-bin/ssdi.cgi?lastname=burchard&firstname=roland&nt=exact (accessed March 15, 2006).
28. Birdseye, "Surveying the Grand Canyon," 21.
29. Citation for Commendable Service, February 2, 1956, Burchard Personnel File.
30. Handwritten notes, n.d., apparently from an interview with Roland Burchard by Otis Marston, box 25, folder 17, Marston Collection.

lands. He spent the remainder of his life in this post, an effective and trusted administrator who frequently advised congressional committees and the secretary of the interior. He died in 1942.[31]

Stabler's biographer noted that "as a Quaker he was peaceful" but "aroused by a worthy cause, he fought openly, squarely and generally successfully with those with whom he disagreed, regardless of whether they were employees, associates, or employers."[32] In other words, he had a lot in common with La Rue, although Stabler had more political acumen and tact. In addition, Stabler "had the ability to obtain and hold the respect and confidence of employees and administrators alike and was the confidant of many in all walks of life with respect to personal problems." Stabler also fought for just compensation for young and underpaid engineers. The Grand Canyon expedition made it into Stabler's memoir, which noted that "he had the distinction of being the one person in the party who was not at some time either upset or thrown off a boat in passing through the rapids. This accomplishment was typical of his tenacity in sticking to a problem."[33]

Frank Dodge was hired as a rodman but ended up as the fifth boatman on the 1923 USGS expedition. After leaving the river at Needles, Dodge continued his itinerant lifestyle, finding work where he could. He probably never married, although he may have at least considered it; in a 1933 letter to Dodge, La Rue, with whom Dodge maintained a friendly correspondence, asked, "Did you marry the girl or did you put an extra blanket on your bed during the cold spell?"[34] Other than perhaps Felix Kominsky, Dodge may have been the most well-liked member of the crew, and those with authority sought to help him for the rest of his life.

Birdseye helped Dodge get a job with a 1924 USGS arctic exploring party as a rodman and boatman.[35] Dodge worked off and on for the USGS for the next two-and-a-half decades, as a recorder, observer, laborer, and engineering aide.[36] In 1927, he was back on the Colorado River, this time as head boatman with the Pathé-Bray movie trip from Green River, Utah, to Hermit Rapid in the Grand Canyon.[37] La Rue, the trip's technical advisor, had recommended

---

31. Hoyt, "Memoir of Herman Stabler," 1643–1645.
32. Ibid.,1644.
33. Ibid.,1643.
34. Eugene La Rue to Frank Dodge, January 27, 1933, box 3, folder 1, La Rue Collection.
35. Dodge, *Saga*, 38.
36. Frank B. Dodge, Personnel Information Record, December, 1941, Dodge Personnel File.
37. Lavender, *River Runners*, 71–75; Dodge, *Saga*, 57–61. That Dodge would work as a whitewater boatman may have surprised Blake, who once asked Dodge if he wanted to take

Dodge to the movie company, because, as Dodge put it, "I was decent to him [on the 1923 trip]."[38] La Rue employed Dodge again in 1935, to oversee the boat work necessary for the Fairchild Aerial Surveys, and they made two river trips from Diamond Creek to Pearce Ferry. It was not until 1937, when hired as the head boatman for the Carnegie Institution-California Institute of Technology Expedition, that Dodge was able to row the entire length of the Grand Canyon.[39] This was the last river trip to run Lava Cliff Rapid before it submerged beneath the rising waters of Lake Mead.[40]

The 1930s found Dodge intermittently working at Lee's Ferry as a stream gager.[41] His autobiography,[42] which includes his account of his time at Lee's Ferry, suggests that he was an alcoholic in the latter half of his life. In 1947, his body crippled by arthritis, Dodge was granted disability retirement from the USGS.[43] He spent much of the rest of his life in hospitals and rest homes, generally broke, and inebriated against his pain.[44] He died in 1965 at his own hand. In 1967, his sister Charlotte wrote a letter begging that Dodge's autobiography, serially written for a USGS publication in 1944, not be distributed. "Many parts of it are fictional and some of the omissions ... have caused years of heartache for his mother and me. Why should old sores be reopened? I've never told anyone that he shot himself late last year about this time."[45]

We respectfully disagree with Dodge's sister. Despite some alleged fictional episodes, Dodge's story speaks strongly of the peripatetic life of a colorful character, mostly lived out in the wildlands

---

the boat past some rocks into fast water. Dodge said he would if Blake didn't think he could make it, but cited his weak wrists. Blake wrote, "I never offered to let him run any swift water again as I did not want to embarrass him." Elwyn Blake to Otis Marston, November 11, 1947, box 21, folder 6, Marston Collection.

38. Frank Dodge to Claude Birdseye, January 17, year unknown, NARA, Record Group 57, Records of the Topographic Division.
39. Dodge, *Saga*, 69–72.
40. Just a few days before the Carnegie-Cal Tech trip ran Lava Cliff Rapid, Buzz Holmstrom ran it on his remarkable solo boat voyage from Green River, Wyoming to Hoover Dam. Vince Welch, Cort Conley, Brad Dimock, *The Doing of the Thing: The Brief, Brilliant Whitewater Career of Buzz Holmstrom* (Flagstaff, Arizona: Fretwater Press, 1998).
41. Reilly, *Lee's Ferry*, 523.
42. Dodge, *Saga*.
43. "Notification of Personnel Action," August 7, 1947, Dodge Personnel File.
44. Otis Marston to "Norah," February 25, 1952, box 51, folder 21, Marston Collection.
45. Charlotte P. Dodge to Dewitt Alexander, November 10, 1967, box 51, folder 21, Marston Collection. California death records indicate that Dodge died in 1965, not 1966; Charlotte Dodge may have had the year wrong in the heading of her letter. California Death Records, RootsWeb.com, http://vitals.rootsweb.com/ca/death/search.cgi?surname=dodge&given=francis (accessed March 6, 2006).

of northern Arizona. Frank Dodge had a vivid and full existence, both in and out of the ocean and on and off of the rivers. He was the one true hero of the 1923 USGS expedition, the only one to have pulled his colleagues from the river when they were in danger. Despite losing his own boat early on, he focused on the safety of his crewmates, particularly the boatmen who often ran rapids with only Dodge standing between a boat flip and a long, perhaps deadly, swim. As he showed with his actions at Upset Rapid, a person like Dodge is someone that all modern river trips would like to have among their crew members. His story deserves to be heard.

The Geologist

Raymond Moore resumed his previous jobs in Lawrence, with the University of Kansas and the Kansas Geological Survey. His diary was filled with geological observations and the types of notes a scientist would keep while in the course of field work; he made no significant written observations on the social interactions on the trip until much later. He published a brief article for a scientific journal[46] and prepared geological reports for La Rue's monograph,[47] despite his dislike for the man.

In addition to his jobs in Kansas, Moore intermittently worked for USGS. One of Moore's biographers noted that while Moore was "the consummate and committed scholar," his personal life was "probably not a very happy one."[48] While not unfeeling, Moore found it difficult to express affection for others. His first marriage, which produced a daughter, ended in divorce; he married Lillian Boggs in 1936. That he was rigid in his outlook is amply demonstrated by an incident from the 1923 expedition, when he scolded Blake for criticizing a bad poem in a magazine; as Blake later noted in a letter to river historian Otis "Dock" Marston," he "told me if I could not do better that I should keep my mouth shut."[49] Marston responded, "According to him [Moore] you must not criticize an egg if you can't lay one."[50]

---

46. Raymond C. Moore, "Geological observations on a traverse through the Grand Canyon of the Colorado," Geological Society of America Bulletin 35 (1924): 85–87.
47. La Rue, *Water Power and Flood Control.*
48. Merriam, "Rock Stars," 18.
49. Elwyn Blake to Otis Marston, December 1, 1947, box 21, folder 6, Marston Collection. In the same letter, Blake details how he proceeded to write a poem, which he showed to Moore. "He read it with apparent interest and handed it back to me with the terse comment: 'That's all right.'"
50. Otis Marston to Elwyn Blake, December 7, 1947, box 21, folder 6, Marston Collection.

Moore had a lengthy and distinguished career as a geologist until he died in 1974 at the age of eighty-two. Even at that time, he was editing his masterpiece, the massive *Treatise on Invertebrate Paleontology*.[51] In 1991, the Kansas Geological Survey honored Moore by taking a Grand Canyon trip on the sixty-eighth anniversary of the 1923 USGS expedition and replicating some of Moore's photographs.[52]

The Junior Boatmen

Elwyn Blake returned to his home in Monticello, carrying with him a passion for rivers that would last the rest of his life. He would go back to the Green River, the place of his first boatman job in 1922, and row from Wyoming to Green River, Utah.[53] He turned down a head boatman position on the Pathé-Bray expedition through Grand Canyon in 1927, citing the cold weather and poor light for photography,[54] although Dodge claimed that Blake demanded too high a salary and "put in some clauses about who the other boatmen should be which I think did not set well."[55] Blake stayed in touch with Emery Kolb and was also a faithful friend to Bert Loper, who was Blake's early mentor on river running.[56] Of Dodge, Blake wrote "... he and I were closer than he and any other man on the '23 trip."[57]

Blake traveled by river at every opportunity, but his life was firmly rooted on land. He married Charlotte Coleman, with whom he had nine children. Blake worked in the newspaper business in Durango, Colorado, and Albuquerque, New Mexico. In 1972, he rode as a passenger through Cataract Canyon, completing his circuit of most of the whitewater reaches in the lower Colorado River system, excluding only Westwater Canyon. He died in 1980.[58]

In his published account, Blake largely kept his feelings about his crewmates to himself. In the late 1940s, Marston sought material for his proposed (but never published) comprehensive history of river travel in Grand Canyon. Marston masterfully goaded Blake

---

51. Merriam, "Rock Stars," 16–18.
52. Baars and Buchanan, *The Canyon Revisited*.
53. Westwood, *Rough-Water Man*, 232–35.
54. Elwyn Blake to Otis Marston, November 11, 1947, box 21, folder 6, Marston Collection.
55. Frank Dodge to Claude Birdseye, January 17, year unknown, NARA, Record Group 57, Records of the Topographic Division.
56. There are many friendly letters between Blake and Loper in box 21, Marston Collection. See also Westwood, *Rough-Water Man*, 233–34.
57. Elwyn Blake to Otis Marston, January 2, 1948, box 21, folder 7, Marston Collection.
58. For more details about Elwyn Blake, see Westwood, *Rough-Water Man*.

for lurid details of the 1923 USGS expedition, focusing mostly on Freeman. Blake responded benignly, so Marston kept prodding until Blake erupted: "As one river rat to another, I think you are full of hooey when you get the idea that I was hard on Freeman in my diary. The fact is, I softened the truth considerably and left out many unpleasant and uncomplimentary matters that irked us most." He noted that at Hance Rapid, a camp rumor had it that Freeman was to be sent out "after his notoriously poor run." Blake demurred from offering his true opinion of Freeman to Birdseye for fear of being dismissed himself.[59]

The friendship between Leigh Lint and Edith Kolb evolved into a romance that went nowhere. Edith sent Leigh beguiling photographs with flirtatious inscriptions. One shows her astride a rearing pinto pony, a few feet from the rim of the Grand Canyon, on which she wrote "This isn't faked Leigh."[60] Both Edith and Lint were young and playing the field. Edith would ultimately marry a Grand Canyon park ranger, Carl Lehnert,[61] while Lint wed his high-school sweetheart, Ruby Rock.[62] In spite of Kolb's apparent fear that his daughter would marry Lint,[63] Leigh and Emery were lifelong friends. Edith had one child, a son, named after her father; she died in 1978.

Lint, whom Birdseye repeatedly commended for having the best runs of the four boatmen, did not return to running rivers. The Birdseye expedition was Lint's fifth and last major run as a young boatman and served as a transition into a career in topography.[64] Undoubtedly encouraged by Birdseye, he worked on a USGS crew surveying lava flows that followed new eruptions in Hawaii. His next assignments found him working throughout the southwestern United States as a junior topographer and engineering aide until 1928.

Lint resigned from USGS in order to pursue topographic engineering studies at the University of Idaho at Moscow, but the Great Depression cut short his educational aspirations. He worked in the field of topography on USGS and US Forest Service (USFS)

---

59. Elwyn Blake to Otis Marston, April 8, 1950, box 21, folder 8, Marston Collection.
60. Unnumbered photograph of Edith Kolb, courtesy of George Lint.
61. Edith Kolb married National Park Service Ranger Carl Lehnert; Suran, "With the Wings of an Angel," chap. 9.
62. George Lint, personal communication.
63. Typewritten note from Dick Gilliland, December 24, 1952, box 111, folder 16, Marston Collection.
64. George Lint, personal communication, March 1, 2006.

survey projects throughout northern California until World War II, when he went east to Virginia to do war mapping. After the war, he returned to California and the Forest Service. Over the next twenty-seven years, he mapped topography for almost all of northern California, including the Hetch Hetchy, Trinity, Mt. Lassen, Sequoia, Kings Canyon, and Modoc Lava Beds projects. After forty-seven years of government service, he retired in 1972, at the age of seventy, from his post as USFS Chief of Surveys and Maps for Region 5 (California).[65] In the early 1970s, the American Society of Photogrammetry recorded commentary from Lint for a soundtrack to an edited version of La Rue's silent movie and tribute to the 1923 USGS expedition.

Lint was a hardworking perfectionist his entire life. He had no patience for fools or politics, which ultimately capped his career at a regional level. Office time was a hardship, as he far preferred field work. His life was his work, which cost him two marriages. He fathered a son with each of his wives; both sons, born fifteen years apart, became military officers. Lint, perhaps having been hampered by the lack of a college degree himself, instilled in his sons the importance of higher education, and both sons obtained postgraduate degrees. Surprisingly, the two sons not only did not meet, but were unaware of each other's existence until after their father's death in 1975.[66]

The Publicist

Freeman raced from Needles to his home and typewriter, and his fingers may have started flying before he had washed all the mud off his body. In November, he received a letter from USGS director George Otis Smith, granting him permission to write an article for the National Geographic Society, noting: "Mr. Birdseye has told me of the excellent spirit in which you did everything possible to help the progress of the party and I wish to convey to you my personal appreciation of your efforts to make the expedition a success."[67] Other members of the crew may have resented his imperious personality and deficient work ethic, but USGS administrators

---

65. George Lint, letter to author, April 28, 2006; Leigh Lint, Application for Federal Employment, November 25, 1946, Lint Personnel File; Leigh Lint, Application for Retirement, June 29, 1972, Lint Personnel File.
66. George Lint, letter to author, April 28, 2006. This paragraph is nearly verbatim from George Lint's letter. Lint's second wife was Helen Long.
67. George O. Smith to Lewis Freeman, November 15, 1923, box 19, folder 33, Marston Collection.

knew why Freeman was on the expedition. Now came the repayment, both to USGS and the crew members who could not restrain expressing their dislike of his presence on the trip.

In addition to his *National Geographic* article,[68] Freeman wrote a three-part article for *Sunset Magazine*,[69] which he gathered into one of his many books, *Down the Grand Canyon*.[70] Both articles and the book were published in 1924. Freeman lectured, illustrating his talks with his photographs as well as those taken by La Rue and Kolb. He repaid those he respected on the expedition, and slighted the others. In his book, he fawned over Birdseye, Stabler, Moore, Burchard, Kominsky (whom he erroneously called "Felix Homs"), and even Roger Birdseye. He allotted only minor attention to La Rue, who most frequently rode on Freeman's boat and swam with him through Separation Rapid. Dodge and the junior boatmen did not even warrant having their first initials in the account, while Frank Word might as well have been elsewhere in the summer of 1923.

In a chapter covering his participation on La Rue's 1922 Glen Canyon trip, Freeman revealed his favoritism for Davis's plan for a high dam in Black Canyon, which he admired for its "impeccable logic."[71] Freeman particularly rankled Lint and Blake by referring to them as Kominsky's "assistants" in the caption of the famous Kolb photograph of the three posing at Diamond Creek with the kitchen kit. This photograph, prominently displayed in his *National Geographic* article and other publications, is one of the iconic images of the expedition, and Blake later acknowledged the slight in the caption was probably turnabout for the treatment Freeman received at their hands.[72] Kolb claimed to avoid the issue: "Knowing Freeman as I do, my time will never be wasted on any of his writings and I have never read his book."[73] Birdseye responded differently: "I am very much pleased with the way you have handled the

---

68. Lewis R. Freeman, "Surveying the Grand Canyon of the Colorado," *The National Geographic Magazine* 45 (1924): 471–530, 547–48.
69. Lewis R. Freeman, "Hell and High Water: Boating Adventures in the Rapids of the World's Deepest River," *Sunset Magazine* (August 1924): 9–13, 57–59; "Hell and High Water: The Colorado Dragon Takes a Bite Out of the Explorers," *Sunset Magazine* (September 1924): 24–27, 62–63; "The Grand Cañon's Roaring Rapids Vanquish the Explorers," *Sunset Magazine* (October 1924): 16–19, 52–56.
70. Freeman, *Down the Grand Canyon*.
71. Freeman, *Down the Grand Canyon*, 198. The book was widely reviewed in newspapers and magazines ranging from the *Arizona Republican* to the *Times of London* and the *Saturday Review of Literature*. See box 70, folder 1, Marston Collection, for clippings of the reviews.
72. Elwyn Blake to Otis Marston, April 2, 1969, box 69, folder 16, Marston Collection.
73. Emery Kolb to Otis Marston, October 17, 1947, box 280, folder 35, Marston Collection.

Blake, Kominsky, and Lint reportedly at Diamond Creek, but probably at Lava Falls. Freeman, in publishing this image in his *National Geographic* article, captioned it "Felix, the cook, and his assistants, Blake (left) and Lint (right)," thus demoting the two junior boatmen.

Emery Kolb Collection, NAU.PH.568-3330, courtesy of the Cline Library, Northern Arizona University.

entire subject and the book has caused a great deal of favorable comment."[74] Birdseye's positive opinion was, of course, the one that mattered.

That Freeman clashed with Kolb, Blake, and Lint was hardly surprising. His world of high society, wealth, and literacy could not have been more different from theirs. Freeman may have viewed Kolb as a small-time operator whose photographic skills were no better (and perhaps worse) than his own. Blake was perhaps stuck in the same category as a small-town journalist. Lint was less than half Freeman's age, heavily imbued with youthful invincibility, and lacking the perspective that might have decreased the personality gap. Freeman's physical limitations probably grated on Lint, the most powerful member of the group, more than on anyone else;

---

74. Claude Birdseye to Lewis Freeman, March 1, 1924, box 70, folder 1, Marston Collection.

those limitations definitely struck a nerve with Blake.[75] One aspect of Freeman's personality, his positive attitude, suggests that he may not have been as bad a boatman as the others claimed; because of his more worldly experiences, he may just not have taken boating through Grand Canyon as seriously as the others.

During the rest of his successful career, Freeman had several additional adventures of the magnitude of the 1923 USGS expedition, some of them involving boats and rivers. His travels took him to the great watercourses of the United States, to South America, and to the Caribbean. He published at least eight more books and scores of articles, and went on numerous lecture tours.[76] He was able to capitalize on the 1923 expedition again in his 1937 compilation entitled *Many Rivers*,[77] which included an account similar to that published in *Down the Grand Canyon*.

Freeman's writing was well received and appears to have been lucrative. His life was not without its difficulties, however; while in Florida during an extended bicycle tour in 1935, he was hit by a car and was seriously injured, which left him disabled for years.[78] In 1955, he was robbed in New York, and possibly injured, and this incident may have contributed to his ensuing senility.[79] Freeman died in 1960 in Pasadena.[80] It says something about the man that his last will and testament not only covered his substantial estate, but also his "books, curios, [and] tennis prizes."[81]

The Cooks

As discussed in chapter 6, Frank Word, upon hearing of the successful conclusion of the expedition, sent a letter to La Rue, which suggests that La Rue was one of the few members of the expedition that Word respected. As shown with scattered comments from the diaries, as well as the gift of a diamond pin to Kominsky at Needles and disregard for Word following the trip, most of the crew did not respect Word or his culinary abilities; ironically, he seemed to have

---

75. Blake, "As I Remember," 166.
76. See box 71, folder 18 of the Marston Collection for further details about Freeman's career.
77. Lewis R. Freeman, *Many Rivers* (New York: Dodd, Mead & Company, 1937): 332–68.
78. "Lewis R. Freeman Hurt: Author is Struck by Auto While Bicycling in Florida," *New York Times*, October 30, 1935 (accessed May 30, 2006).
79. Otis Marston to Roland Burchard, September 9, 1962, box 25, folder 17, Marston Collection.
80. Friman, "Lewis Ransome Freeman," 143.
81. Estate of Freeman (1965) 238 CA2d 486, http://online.ceb.com/calcases/CA2/238CA2d486.htm (accessed October 13, 2005).

had a good deal of support from Kolb, whom Word greatly disliked. The letter to La Rue is the last time Frank Word is heard of from the perspective of the 1923 USGS expedition. He continued to work as a cook in Los Angeles through at least the middle years of the 1920s.[82]

The other cook, Felix Kominsky, was fondly remembered. In 1924, La Rue tried to hire him for another Colorado River trip, this time traveling two hundred miles downstream from Pearce Ferry. Roger Birdseye, living in Flagstaff, wrote to La Rue:

> Felix Koms has been a bird of passage since the end of last year's trip but so far as I am able to trace him he is now cooking at a lumber camp at McNary[83] ... I hope by tomorrow to have a more definite address and will then write him immediately, attempting to put the bee in his bonnet. I understand that he has taken unto his busom [sic] a wife, which may complicate matters somewhat. Otherwise I am sure he would be glad to go, provided he is not too firmly anchored. The same interests that own most of the Cooley camps have recently bought the Flagstaff Lumber Company and I may be able to arrange with the Powers That Be that he get his job back if that factor worries him. Do you want me to quote him terms? If so, will they be the same—transportation and $150 per month?[84]

Kominsky was either unavailable or did not respond, and, like Word, he slipped into obscurity.

Emery Kolb, Claude Birdseye, and USGS

Back at his photographic studio at the South Rim, Kolb began the lengthy process of processing his films and negatives, which he had obtained despite Birdseye's opposition to his photographic aspirations on the expedition. He cut some of his new film footage into the *Grand Canyon Film Show*, the bread-and-butter motion picture projected daily at the Kolb Studio. This show became the longest-running movie in motion-picture history.[85] Emery then began

---

82. Word is traceable in the 1924–27 issues of the Los Angeles City Directory. For 1924, see p. 2,394; for 1925, see p. 2,041; for 1926, see p. 2,129; for 1927, see p. 2,095.
83. McNary is located southeast of Flagstaff, and was a lumbering town in 1923.
84. Roger Birdseye to Eugene La Rue, August 3, 1924, box 2, folder 2, La Rue Collection.
85. The longevity of the film is from Pedersen, *Emery Kolb*, 6. By 1969, Kolb estimated that the film had been shown some 60,000 times. Suran, "With the Wings of an Angel," chap. 11.

his long seditious war with USGS, believing his rights as a small business owner were usurped by an agency of the federal government.

Kolb and Birdseye exchanged letters and images, but these never resolved Kolb's anger. He was further inflamed when the National Geographic Society declined his offer for a lecture, as they had already made arrangements for a tour with Birdseye. Birdseye wrote Kolb that he had turned down other lectures so as not to compete with Kolb, and that he had tried to reject the Society's offer; this attempted placation did not assuage Kolb. Insults piled on slights. Correctly or not, at least one version of the film of the trip footage released by the USGS lists La Rue as the cinematographer, not Kolb. In all likelihood, this attribution was correct, but as the producer of the *Grand Canyon Film Show* and self-proclaimed czar of all images Grand Canyon, Kolb was outraged.

In June 1925, Kolb composed a seventeen-page document outlining his complaints against USGS, largely directed at Birdseye, but also at Freeman, La Rue, Moore, and Stabler. In Kolb's world view, the USGS had "goldbricked a citizen" and he had "been grossly wronged in the non-fulfillment of promises offered to me as inducement and as part remuneration for my services."[86] Kolb attempted to use his friends in high places; he sent the letter to his friend Senator Ralph Cameron from Arizona. He requested that Cameron submit the document to a Senate committee investigating wrongdoing in governmental departments.

Cameron, the junior senator, was much more politically savvy than Kolb. Kolb, the small businessman occupying property within a national park, was vulnerable and stood more to lose than gain with his complaint. Cameron strongly advised Kolb against publishing his diatribe, suggesting that doing so would only "arouse the vengeance" of the Department of the Interior, not a wise thing to do while a national park concessionaire.[87] Kolb appears to have taken Cameron's advice, albeit begrudgingly.

Kolb nursed his anger against USGS, probably for the rest of his life. His feelings toward Birdseye were ambivalent; he could hardly forget the calming influence that Birdseye had over the expedition, mostly oriented toward defusing the rancor that Kolb caused. In 1956, he wrote to Burchard: "I always regretted the little unpleasantness I had with Colonel Birdseye in not keeping several

---

86. Emery C. Kolb, untitled typescript, June 18, 1925, box 15, folder 1769, Kolb Collection.
87. Ralph H. Cameron to Emery Kolb, July 17, 1925, box 5, folder 750, Kolb Collection.

promises with me, and regardless of this I could never feel bitter toward him. I really thought he was a fine fellow and liked him. I believe it was just a habit of his, that when he wanted something, he would make one any kind of a promise, and then forget it or go back on it, if it didn't suit to fulfill it."[88] In other words, Kolb believed that Birdseye, first and foremost, was a bureaucrat with a mission to fulfill; his saving grace was that he was a genial bureaucrat cursed with a fleeting memory of promises. Of course, in Kolb's version, all those forgotten promises were ones that would have benefited Kolb. We find no evidence that Birdseye actually made the promises that Kolb claimed were made.

As a glaring example, Kolb offered Burchard this snippet from his 1924 visit to Washington, when he was a guest at the Birdseye home. After a disagreement and out of Claude's presence, Kolb said to Mrs. Birdseye, "Here I am in Washington to see the Colonel about some promises he made me. Mrs. Birdseye straightened up and looked at me. I thought I was going to be rebuked, but instead she said, 'Mr. Kolb. If Colonel made promises he did not keep, you keep after him and see that he keeps them. I don't like that,' she said."[89] Kolb took this to mean that Claude Birdseye's propensity for forgotten promises was a fault known at least to his wife.

Not receiving satisfaction from Birdseye, Kolb still wanted something from USGS to at least show the agency's appreciation of what he thought his contributions were, despite USGS pointedly mentioning his contributions in the 1923 expedition press release. In the same 1956 letter to Burchard, Kolb asked for something in writing acknowledging Kolb's activities on the 1923 USGS expedition. Kolb felt he had "practically nothing to show in writing the part I played."[90] It was not a fair complaint: Kolb's role had been prominently discussed in the publications by Birdseye, Moore, and Freeman, and Kolb adamantly refused to read the latter's book. In addition, Birdseye, in his letters and lectures, gave profuse credit to Kolb. To again calm the volatile Kolb, Burchard complied; much to Kolb's disappointment, the letter was not on USGS letterhead, although it was published as a letter to the editor in the *Arizona Republic*.[91]

---

88. Emery Kolb to Roland Burchard, September 16, 1956, box 111, folder 20, Marston Collection.
89. Ibid.
90. Ibid.
91. Roland Burchard to "Whom it may concern," undated but known to be October 19, 1956, box 15, folder 1814, Kolb Collection. Kolb's lament that it was not on USGS

Kolb was a fixture at the South Rim, and his paranoia about government intervention to the contrary, Grand Canyon National Park allowed him to ply his trade for the rest of his life. He continued occasional river excursions, and both Emery and Ellsworth were sometimes called upon to aid in rescues. In 1927, when the Pathé-Bray motion-picture expedition (of which La Rue was a member) was reported lost, Emery declined participation in the ensuing search: his suspicion that the disappearance was a publicity stunt was accurate.[92] The brothers did aid in the search for the genuinely missing (and never found) honeymoon couple, Glen and Bessie Hyde, the following year.[93] In 1938, he ran the lower part of the Grand Canyon, from the Bright Angel Trail junction down to Lake Mead with Norman Nevills. Two years later, Emery rode with Barry Goldwater, who later became a senator from Arizona, through Bright Angel Rapid for the first-ever radio broadcast from the Grand Canyon, for a *Ripley's Believe it or Not* segment.[94]

In the mid-1950s, Kolb floated an idea to his old pal Leigh Lint: would Lint run Grand Canyon with him again? Lint declined, citing a bad back.[95] Kolb's wife, Blanche, died in 1960, the same year that Ellsworth passed on.[96] These losses may have slowed Emery, but they did not stop him. Kolb had one more opportunity to run a few of the canyon's rapids in 1974. He was flown by helicopter to the Little Colorado River-Colorado River confluence to join a motorized river trip that included river historian Dock Marston. Emery, ninety-three, and frail, rode as far as Crystal Rapid, then flew back out.[97] He would die two years later in a Flagstaff hospital, having outlived every other member of the 1923 expedition except Elwyn Blake, who was fifteen years his junior.

Kolb had come to accept that his beloved studio would be torn down after his death, as the National Park Service had made it repeatedly and abundantly clear that that was their goal ever since the park was established in 1919. In 1962, the Park presented Kolb with a contract stating that the studio would become government property upon his death and would be demolished. Kolb balked,

---

stationery and the date of the letter, appear in association with a copy in box 25, folder 17, Marston Collection.
92. Suran, "With the Wings of an Angel," chap. 9.
93. Dimock, *Sunk Without a Sound*.
94. Suran, "With the Wings of an Angel," chap. 9.
95. Ibid.
96. Suran, "With the Wings of an Angel," chap. 11.
97. Suran, "With the Wings of an Angel," chap. 9.

Sketch of Eugene C. La Rue by R. C. Moore. Edith Kolb diary, Emery Kolb Collection, box 14, folder 1756.

Courtesy of the Cline Library Northern Arizona University.

but had no recourse other than to sign.[98] In 1966, the National Historic Preservation Act was passed, which specifically provided protections for buildings such as the Kolb Studio. Kolb, through his longevity, had the last laugh: after his death, the building was listed on the National Register of Historic Places and was renovated by the Grand Canyon Association (GCA). It now houses a bookstore and exhibit space for GCA, an organization that supports education, interpretation, and visitor services within the park.[99]

Despite his protestations to Birdseye, USGS usurped none of Kolb's business nor his reputation. Emery Kolb is a more-recognized figure in Colorado River photography than either of his rivals, Freeman or La Rue, or even the Fox News cameraman, J. P. Shurtliff. Freeman and La Rue's photographs are housed in the obscure U.S. Geological Survey Photograph Library and are seldom viewed, while Kolb's photographs repeatedly are used to illustrate the daring-do of early Grand Canyon adventurers. Kolb remains revered in Grand Canyon river history as well as its landscape photography.[100] Colorful, pugnacious, tough, and enduring: that was Emery Kolb.

---

98. Ibid.
99. Grand Canyon Association, http://www.grandcanyon.org/ (accessed March 7, 2006).
100. Lavender, *River Runners*; Webb, Belnap, and Weisheit, *Cataract Canyon*.

The Fate of the Hydraulic Engineer

For Eugene La Rue, the Grand Canyon expedition was merely one more piece of critical research, another data set to add to his growing arsenal of information on the Colorado River. The fact that he seemed to have alienated nearly everyone on the expedition did not impress or concern him. Self-centered and self-absorbed, La Rue pushed ahead, ignoring the subtle and not-so-subtle warnings about his personality that he received while on the river, even from his wife. He would ultimately pay for ignoring those warnings.

La Rue eagerly made his case for his high dam upstream from Lee's Ferry. He played to the general public, and in so doing, he had to know he was circumventing the still-maturing USGS. The more public his stature, the more attention La Rue received from his less-than-sympathetic supervisors. The Department of the Interior took notice of the growing feud between La Rue and the Bureau of Reclamation and discussed the matter with Director Nathan Grover. That attention was not well received, and USGS administrators may well have retained a collective memory of the funding difficulties that arose from Powell's advocacy of Colorado River water development in the 1890s.

La Rue's nemesis, Arthur Powell Davis, was now out of the picture. Davis was forced to resign from the Reclamation Service before the 1923 USGS expedition. How La Rue felt about Davis's downfall is lost to history, but one cannot help but believe that he would have been very pleased: he could now push his water-development agenda with his main antagonist sidelined. He vociferously advocated his damsite and his water-development plan, without the official consent, support, or approval of USGS or the Department of the Interior. To provide scientific justification, he worked hard to produce his seminal monograph on the Colorado River.

On October 4, 1923, La Rue received a letter from Grover, asking him to "please take especial care not to discuss or even mention controversial matters," particularly as they related to the "pact" [Colorado River Compact].[101] La Rue refused to comply. Later that month, in a letter to his colleague G. E. P. Smith, an irrigation engineer at the University of Arizona, La Rue wrote "I have just read your article ... 'Colorado River Projects—Which One First.' I wish I could discuss these matters with you personally for you have gone

---

101. Nathan Grover to Eugene La Rue, October 4, 1923, box 2, folder 1, La Rue Collection.

wrong in 57 different places." La Rue arrogantly closed the letter, "I believe it would help clear the atmosphere if the politicians as well as engineers would postpone discussion of the Colorado River projects until the engineering facts are made available [La Rue's forthcoming report]."[102]

La Rue's monograph was published two years to the day after the 1923 USGS expedition landed at Needles. The report, entitled *Water Power and Flood Control of Colorado River below Green River, Utah*,[103] is a masterpiece in scientific exposition that would have made John Wesley Powell proud. La Rue's introduction is telling: "The purpose of this report is to present the facts regarding available water supply and all known dam sites on Colorado River between Cataract Canyon, Utah and Parker, Ariz., and to show the relative value of these dam sites. To determine the relative value of the dam sites, *a comprehensive plan of development* [emphasis ours] for Colorado River below the mouth of Green River is presented that will provide for the maximum practicable utilization of the potential power, maximum preservation of water for irrigation, effective elimination of the flood menace, and adequate solution of the silt problem."[104]

La Rue's plan called for thirteen dams, including a high dam in Glen Canyon as well as a smaller one in Mohave Canyon near Needles (see illustration on page 274). Specifically, he proposed "a reservoir site below the mouth of Green River and above Grand Canyon to regulate the flow in the interest of power development, and another reservoir site below all the large power sites to prevent floods on the lower river and re-regulate the flow in the interest of irrigation."[105] A high dam in Glen Canyon, in one sense, would make political sense; its location was about five-and-one-half miles upstream from Compact Point, the dividing point between the upper and lower basin states. As such, La Rue's dam would regulate water transfers between the two groups of states.

A series of smaller dams would create a "heel-to-toe staircase of impoundments"[106] that would generate hydropower, a plan emulated, if not copied, for the Colorado River Storage Project and its

---

102. Eugene La Rue to G. E. P. Smith, box 4, folder 15, Smith Collection MS 280, University of Arizona Special Collections, Tucson, AZ.
103. La Rue, *Water Power and Flood Control*.
104. Ibid., 9.
105. Ibid., 10.
106. Langbein, "L'Affaire LaRue" (1983):43.

"cash-register dams" proposed in the 1950s.[107] In Black and Boulder canyons, La Rue proposed three such dams, instead of the one high dam advocated by the Fall-Davis report and Swing-Johnson bills. While power generation might be the prime objective, water conservation would drive the size and location of dams. As Powell had advocated four decades previously, minimal reservoir surface area means minimal evaporative losses and more available water. The difference was that Powell would have built dams closer to the cooler headwaters; La Rue's dams were either on the hotter Colorado Plateau or in the even hotter Basin and Range Province.

La Rue stressed that his was a preliminary plan, but he emphasized the need for it to become comprehensive. Grover's introduction to La Rue's treatise carefully stated, "The Geological Survey is not attempting to promote any particular project ... It recognizes ... that the final choice of any project will represent a compromise of conflicting interests ..."[108] But this warning was little more than a figurative wet blanket that did little to mask La Rue's convictions. As investigations continued, La Rue's plan grew to include a gravity aqueduct running from a dam at Bridge Canyon in western Grand Canyon to Los Angeles, which, because of the intervening topography, constituted a major engineering design problem. He continued his program of self promotion, emphasizing media opinion in the region that could most benefit: Southern California.

La Rue sent a copy of his plan to *Los Angeles Times* editor Harry Chandler for his considered opinion.[109] While the *Times* supported La Rue's plans, owing to its ulterior motives in Mexico, larger water powers opposed him. William Mulholland, the chief engineer of the Los Angeles Water Department, advocated construction of a much shorter aqueduct with an intake somewhere near Needles, with the water being pumped to an elevation sufficient for mostly gravity-driven delivery to Southern California. Hydroelectric power, preferably from a high dam in Boulder Canyon, would be used to pump the water out of the river and through the Mojave Desert. In October 1925, Mulholland ridiculed La Rue's aqueduct plan in a special session of the U.S. Senate Committee on Irrigation and Reclamation in Los Angeles. The *Times*, in a front-page story, quoted Mulholland as saying "I will not say it can't be done, but

---

107. Reisner, *Cadillac Desert*.
108. La Rue, *Water Power and Flood Control*, 7–8.
109. Eugene La Rue to Harry Chandler, January 5, 1926, box 2, folder 4, La Rue Collection.

it would be so expensive that there isn't that much money in the world."[110] La Rue had support from the state of Arizona, for which he worked as a consulting engineer. At the time, Arizona, it would seem, opposed anything concerning the Colorado River that was supported by California.

In December 1925, La Rue finally got his day in court when he testified before the Senate Committee on Irrigation and Reclamation in Washington, D.C. He had one last chance to make his lone voice heard over the cacophony of support for the Boulder Dam project. La Rue started the meeting with a statement of his qualifications and an extended discussion of his monograph. Senator Wesley Jones of Washington repeatedly interrupted, prompting La Rue to blurt out: "May I finish, please? It will take only about 10 minutes ..." He concluded with the statement: "Gentlemen of the committee, I offer these recommendations as my own, not speaking for the Geological Survey of the Department of the Interior."[111]

La Rue's personality, so revealed during the 1923 USGS expedition, came out in the testimony. At one point, he stated: "... with respect to a gravity water supply for the city of Los Angeles. I have prepared that since I was attacked, and this was just recently. I mean, after I was bitterly attacked in Los Angeles I got busy to find out what was wrong." Senator Charles McNary, the committee chairman, probed La Rue's statements forcefully. Senator Hiram Johnson of California steadily picked at the weaknesses in La Rue's plan and personality, lecturing him at times about the legislative process. La Rue was grilled for four hours, at times offering testy replies to simple questions. This time, not even the Arizona senators, both of whom were on the committee, came to his defense.

La Rue's testimony was filled with statements that we now know to be untrue. He claimed that a dam of the size proposed for Black Canyon was technically infeasible. He stated: "If you construct the Boulder Canyon Dam ... you will have regulated the river and increased the low-water flow to such an extent that it will be far beyond the needs for irrigation in this country for many years to come" and, therefore, would benefit Mexico more than the United States. For his own proposed dam sites, he could offer no data on stability of the rock foundations; drilling would have been necessary

---

110. "Mulholland in Plea for High Dam on Colorado," *Los Angeles Times*, October 28, 1925 (accessed March 7, 2006).

111. Senate Committee on Irrigation and Reclamation, *Colorado River Basin: Hearings on Senate Resolution 320*, 68th Cong., 2nd sess., December 9, 1925, 531–97.

to obtain that, and Southern California Edison did not provide sufficient funding for such work. However, the better-endowed Reclamation had foundation data for the Boulder Canyon site. He had to admit that his work was funded by Southern California Edison, which Johnson noted could bias La Rue toward their anti-Boulder Canyon position.[112]

La Rue countered the opposition as best he could, claiming that Reclamation's plans were wasteful of both money and water. "Mr. La Rue expressed himself as being unalterably opposed to a high dam at Boulder Canyon. He asserted this site ... is one of the least desirable of some forty-seven potential dam sites."[113] Ultimately, La Rue was not persuasive. It was his last moment in the national spotlight and it was a bitter defeat.

The next day, Secretary of Commerce Herbert Hoover held forth to the committee, emphasizing the points made in the Fall-Davis report. Arthur Powell Davis might have left the spotlight, but powerful people stepped in to carry his torch. Hoover stressed that the proximity of the enormous Los Angeles power market to Boulder Canyon made it the most desirable site for the first dam.[114] The nearby railhead in Las Vegas, the presence of major road access, and the local availability of the raw materials for concrete all pointed toward a high dam straddling the Nevada-Arizona border. Political expediency would trump science, and Reclamation scored its first major victory in the war over western water development.

At least initially, La Rue's plan was rejected, although the determined engineer would fight to keep it alive through the mid-1920s. He wrote to the Senate committee asking for an opportunity to cross-examine a former Reclamation engineer who had declared La Rue's plan unsound. The request was denied; the hearings had ended. He asked the president of First National Bank in Los Angeles to meet with him in order to "present my views on the Colorado River matter."[115] He took potshots at Mulholland at every opportunity; in a letter to his brother, La Rue wrote, "I have been opposing him publicly. We are a long ways apart and I am certain

---

112. Senate Committee on Irrigation and Reclamation, *Colorado River Basin: Hearings before the Committee on Irrigation and Reclamation*, 68th Cong., 2nd sess., 1925, 531–97.
113. Kyle D. Palmer, "Clash over Colorado," *Los Angeles Times*, December 10, 1925 (accessed March 7, 2006).
114. Langbein, "L'Affaire LaRue" (1983):44.
115. Eugene La Rue to Henry M. Robinson, February 4, 1926, box 2, folder 4, La Rue Collection.

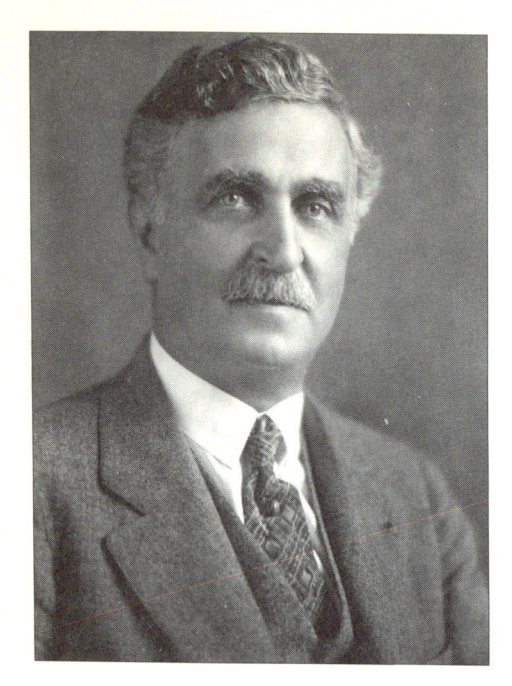

George Otis Smith, fourth director of the U.S. Geological Survey. He ultimately caused the resignation of Eugene C. La Rue, an action that the *Los Angeles Times* deemed to be "muzzling" of the 23-year veteran hydraulic engineer.

Photographer unknown, Portraits 5, courtesy of the U.S. Geological Survey Photograph Library.

that one or the other is wrong."[116] In response to an April 1926 letter from Secretary of the Interior Hubert Work, who said he was sorry he had not had more of an opportunity to speak with him at a recent meeting both men had attended, La Rue replied, "Will you permit me to say that it is my opinion that if an attempt is made to build a dam on Colorado River at the Boulder Canyon or Black Canyon dam site the Bureau of Reclamation will be responsible for having made the most gigantic engineering blunder that has ever been made in this country."[117] History considers La Rue's statement to be the blunder.

Two months later, La Rue received an ominous telegram from USGS Director George Smith: "Criticisms and complaints of your public utterances and even private conversations have become so frequent that only practicable method to defend Survey and yourself from charge of opposing adopted policy is silence Stop In spite of my faith that your intentions are of best in which Secretary Work joins me I must disapprove any further discussions by you of Boulder Dam or related subjects Stop Surveys full and impartial exposition of engineering facts in your published report is our part in this controversy which has now reached situation so tense

---

116. Eugene La Rue to Scott La Rue, February 8, 1926, box 2, folder 4, La Rue Collection.
117. Eugene La Rue to Hubert Work, May 22, 1926, box 2, folder 4, La Rue Collection.

that misunderstanding is sure to follow any further contribution from you."[118]

La Rue chafed under the "muzzle" order, as he termed it, and he began to realize that his USGS career was over. He started to explore other opportunities. He found one in an unlikely place: as a consultant on another Grand Canyon river trip. The Pathé-Bray Company of Hollywood wanted to make a river movie in order to take advantage of the public interest in the Colorado River. Despite his extreme aversion to whitewater, expressed repeatedly during the 1923 Grand Canyon expedition, La Rue committed to Pathé-Bray. He thought a furlough from his government job might be acceptable to USGS. Again, he was wrong.

On July 4, 1927, La Rue made his proposal to Grover. He wired a telegram, "Have accepted job with Bray Pictures Corporation request six months furlough beginning July twenty six if furlough not granted accept my resignation effective same date." Hand-written at the bottom of the telegram is a note from Grover, "Resignation recommended for acceptance. Effective at the close of July 25, 1927."[119] La Rue had gambled his federal government job and now had no choice: he resigned.

La Rue's departure from the USGS did not go unnoticed. The *Los Angeles Times* reported, "Because he was 'muzzled' when he attempted to point out defects in the Boulder Canyon Dam project E. C. LaRue has resigned his position as head of the Pasadena branch of the United States Geological Survey after twenty-three years of government service ... The government silenced the engineer shortly after the Washington hearing, he says, and he feels that, under these conditions, his usefulness is near an end."[120] Insult was added to injury: La Rue's personnel file contains a letter from a Pasadena Investment Securities worker named Benjamin Fenton, congratulating Hubert Work for "relieving the government" of La Rue's services. "Let us hope that Mr. La Rue's successor will give publicity to the Government's plans and views and keep his private opinions strictly to himself."[121]

Approval of La Rue's resignation was not universal. USGS Arizona District Chief William E. Dickinson expressed an opposing sentiment when he wrote to La Rue, "Your resignation ... comes as more

---

118. George O. Smith to Eugene La Rue, June 10, 1926, box 2, folder 4, La Rue Collection.
119. Eugene La Rue to Nathan Grover, July 4, 1927, La Rue Personnel File.
120. "Colorado Dam Expert Resigns," *Los Angeles Times*, July 7, 1923 (accessed March 7, 2006).
121. Benjamin W. Fenton to Hubert Work, July 11, 1927, La Rue Personnel File.

a disappointment than as a surprise. A disappointment because it makes it just that much more apparent that a man cannot really get there in the Service and keep his individuality. Through what contact I have had with you and your work in the Survey, I have received much inspiration and encouragement, and it is with a feeling of personal loss that I contemplate your separation from the organization."[122] Dickinson's comments foreshadow the bureau's policy, enforced for many decades after La Rue's resignation, of an official policy review of all publications as well as strong disapproval of any advocacy on controversial issues.

For La Rue, the decision to leave USGS must have been painful and awkward. He continued to have financial problems, so the resignation had to have taken a toll on his quality of life. On the one side, he was absolutely certain about the superiority of his proposal and felt morally obligated to stop the opposition. "The whole Boulder Dam scheme is so rotten that it almost makes me sick at my stomach when I think of it," he wrote in 1919, at the height of his animosity towards Arthur Davis.[123] And even though he was not always happy with his work, it had been the core of his existence. The press release that coincided with publication of his 1925 Colorado River report noted that La Rue "... has made boat trips aggregating nearly 2,000 miles along the [Colorado] river and its major tributaries ..."[124] In 1928, La Rue wrote to a fellow engineer, "The greatest change in my daily routine of life came when I resigned after 23 years in the service of Uncle Sam. Even when I got married I took my wife on a field trip through Idaho and called it a honeymoon."[125]

In the view of prominent USGS hydrologist Walter Langbein, La Rue had been guided by "the principle of engineering determinism—the advocacy of a single best plan."[126] La Rue clearly was influenced by John Wesley Powell's big science thinking: instead of a piecemeal approach to water development in the West, as represented by the Boulder Dam Project, the best idea was to consider the problem in its entirety. But La Rue's steadfast belief that

---

122. W. E. Dickinson to Eugene La Rue, July 15, 1927, box 2, folder 5, La Rue Collection.
123. Eugene La Rue to George Maxwell, January 11, 1929, box 2, folder 6, La Rue Collection.
124. "Department of the Interior, Memorandum for the Press: Engineering Report on Colorado River," October 19, 1925. NARA, Record Group 57, Records of the Topographic Division.
125. Eugene La Rue to Jack Savage, May 29, 1928, box 2, folder 6, La Rue Collection.
126. Langbein, "L'Affaire LaRue" (1983): 39.

his plan was the right one, followed by his resolute advocacy of his plan, not only led to his resignation, but Langbein claimed that La Rue's advocacy altered the way the Water Resources Branch did business.[127] The branch became mostly a data-collection agency, one that shied away from taking any particular policy stand but instead dealt solely with scientific issues relevant to policy decisions. La Rue's story stresses that when scientific arguments clash with policy, policy may well win.

La Rue indeed advised the Pathé-Bray Film Company in their aborted effort to film a movie about river running on the Colorado River. Imagine how Emery Kolb must have reacted to this, knowing that yet another interloper was making movies about his river, and aided by La Rue and guided by Dodge, to boot! But the "Pride of the Colorado" became the "Bride of the Colorado," and then became discarded footage on the cutting room floor. The river trip, which could hardly be called an expedition, is now best known for its contrived reports of the crew being "lost" in the canyons upstream from Lee's Ferry.[128] Dodge, in a lengthy letter composed to Birdseye sometime after the expedition, wrote, "La Rue has improved some but in many respects is the same. He just can't see himself. On the trip he was a figure head & did nothing except come out in the papers which I suppose was the main purpose."[129]

After the conclusion of the Pathé-Bray debacle, La Rue returned to his scientific career. He formed an engineering consulting firm and continued his battles as a private citizen. When construction of Hoover Dam became inevitable, he still championed his gravity aqueduct plan, which hinged around construction of the Bridge Canyon Dam. Neither came to pass. The remainder of La Rue's life was not easy. His engineering firm failed, and during the Depression, he lost his home. In 1932, he lamented in a letter to Frank Dodge, "Wish I could have a pay job on the river again."[130] In spite of his very real reservations about whitewater, La Rue was drawn to rivers. His fellow travelers on the 1923 expedition often commented about his fears, yet La Rue devoted much of his life to rivers, traveling thousands of miles along the Colorado and Green rivers, much of it involving whitewater.

---

127. Langbein, "L'Affaire LaRue" (1975): 6.
128. Lavender, *River Runners*, 66.
129. Frank Dodge to Claude Birdseye, January 17, year unknown, NARA, Record Group 57, Records of the Topographic Division.
130. Eugene La Rue to Frank Dodge, September 6, 1932, box 3, folder 1, La Rue Collection.

La Rue may have felt at least a bit exonerated when, in 1946, more than a decade after the completion of Hoover Dam and long after the departure of William Mulholland, the City of Los Angeles espoused the building of Bridge Canyon Dam. "... if the dam is built," reported the October 25 edition of the *Los Angeles Times*, "La Rue should receive due credit for discovering the possibilities."[131] There no longer was a need for La Rue's most controversial proposal, the aqueduct from Bridge Canyon to Los Angeles: the Colorado River Aqueduct, transporting water pumped from Lake Havasu, was already under construction. La Rue died of a heart attack five months later at age sixty-seven.[132]

Eugene C. La Rue fell on his own figurative sword, one forged from his hydrological view of the Colorado River and honed by his abrasive and single-minded personality. His nemesis, Arthur Powell Davis, preceded La Rue with his own fall from grace several years previously. Davis ultimately took a position as chief engineer of a utility district in San Francisco, and later worked in the Union of Soviet Socialist Republics as an irrigation consultant. In 1933, Davis achieved a level of government exoneration when another secretary of the interior, Harold Ickes, asked him to serve as a consultant on the Boulder Dam project. Davis declined due to his failing health; he died a month later.[133]

What If

On September 30, 1935, Hoover Dam was dedicated.[134] Parker Dam, which impounds Lake Havasu, was completed in 1938;[135] intakes for both the Colorado River Aqueduct, which transports water to the Coachella Valley and Los Angeles basin, and the Central Arizona Project, which transports water to central and southern Arizona, withdraw their allocations from this water body. Davis Dam, named for Arthur Powell Davis, was completed in 1950.[136] It is essentially a hydropower structure, and Lake Mohave serves

---

131. "Hoover Dam Power is Now a Decade Old," *Los Angeles Times*, October 25, 1946 (accessed March 7, 2006).
132. "Memoir: Eugene Clyde La Rue," *Water Resources Bulletin* (May 10, 1947): 79.
133. Pisani, *Water and American Government*, 137–39; Charles A. Bissell and F. E. Weymouth, "Memoirs of Deceased Members: Arthur Powell Davis, Past-President, Am. Soc. C.E.," *Transactions of the American Society of Civil Engineers* 100 (1035): 1582–91.
134. Stevens, *Hoover Dam*, 243–44.
135. U.S. Bureau of Reclamation, Dams, Projects and Powerplants: Parker Dam, www.usbr.gov/dataweb/dams/az10312.htm (accessed November 4, 2005).
136. U.S. Bureau of Reclamation, Dams, Projects and Powerplants: Davis Dam, www.usbr.gov/dataweb/dams/az10309.htm (accessed November 4, 2005).

primarily as additional storage and a re-regulation impoundment that smoothes out the diurnal releases of the Hoover Dam power plant. Combined with Laguna and Imperial Dams just upstream from Yuma; the small diversion structures, Headgate Rock Dam and Palo Verde Dam, between Parker Dam and Blythe, California; and Morelos Dam, a Mexican diversion structure downstream from Yuma, the lower Colorado River became fully regulated and is mostly diverted for irrigation and domestic consumption.

The Bureau of Reclamation was not finished with the Colorado River. The Colorado River Storage Project, passed by Congress in 1956, became Reclamation's version of La Rue's plan, albeit scaled back owing to the onset of the environmental movement.[137] David Brower, a now-legendary environmentalist, led Reclamation to its Waterloo when the agency proposed the Echo Park Dam, on the Green River in Dinosaur National Monument and upstream from the reaches that La Rue studied in such detail. Environmentalists essentially gave up the Glen Canyon dam site to save Echo Park, a concession Brower rued until the day he died.[138] Glen Canyon Dam, completed in 1963, impounds the Colorado River a little more than ten miles upstream from La Rue's high dam site. Instead of dynamiting the canyon walls and packing the rubble with sand and clay, Reclamation plugged the canyon with concrete in the form of a thick-arch dam. Unbeknownst to La Rue, his foundation was on unstable Chinle Formation—the senators were correct in criticizing his lack of foundation analyses—and Reclamation built Glen Canyon Dam on a solid yet leaky foundation of Navajo Sandstone.

Arizona doggedly pursued dam sites proposed by Girand and La Rue through the twentieth century. From the early 1940s through the early 1950s, Reclamation engineers conducted foundation surveys at several dam sites in Marble Canyon as well as between Gneiss and Bridge canyons in western Grand Canyon. Reclamation promoted the structures, with support from the Arizona Legislature, until they were defeated by an environmental coalition led by David Brower with the support of environmentalist, author, and Grand Canyon boatman Martin Litton.[139] Brower was not to concede

---

137. Reisner, *Cadillac Desert.*
138. Byron E. Pearson: *Still the Wild River Runs: Congress, The Sierra Club, and the Fight to Save Grand Canyon* (Tucson: University of Arizona Press, 2002),35.
139. Pearson, *Still the Wild River Runs.* Pearson argues that the role played by the Sierra Club and other environmental organizations in actually stopping construction of Grand Canyon dams was less instrumental than generally believed, and that the decision was more one of political pragmatism.

another dam site, and his public-relations coup responded to the glorification of the potential beauty of a lake in Grand Canyon: "Should we also flood the Sistine Chapel so tourists can get nearer the ceiling?" Without Reclamation as a player, representatives from Arizona continued to press for a Grand Canyon dam site, periodically introducing resolutions until at least 1981.[140] The era of Colorado River dam construction, at least the major structures on the Colorado Plateau, ended.

In some respects, La Rue's hydrology has proven correct even if some of his vehemently advocated details were not. There is not enough water in the Colorado River watershed to meet demand, a point made abundantly clear by the potential shortages caused by early twenty-first century drought. That drought underscored the wisdom of a comprehensive plan; the current system of regulation, more patchwork than comprehensive, was sufficient to buffer an extreme, three-year drought without missing any of its delivery commitments. In 2006, water allocations, which are not fully extracted, exceed water production by perhaps two million acre-feet. Water evaporates in enormous volumes from the massive surface of Lake Mead, the reservoir created by Hoover Dam, and Lake Powell, the reservoir impounded by Glen Canyon Dam. As noted by hydrologist Langbein, La Rue's three dams would have a combined surface area of 28,000 acres while Lake Mead alone has a surface area of 138,000 acres.[141] As he predicted, less water loss to evaporation would have occurred had La Rue gotten his plan through Congress.

In the illustration on page 275, we compare the flow regulation below the Confluence of the Green and Colorado rivers according to La Rue's plan[142] versus that implemented by Reclamation. From Green River, Utah, to Parker, Arizona, hardly any free-flowing reaches would remain, replaced by a series of lakes that back up from one dam to another. Ironically, La Rue's plan would have spared Dark Canyon Rapid, one of the major rapids he faced twice as a river runner and one ultimately inundated by Lake Powell. His plan would have generated far more hydropower than the existing network of dams; now, considerable potential hydropower is consumed by whitewater in Grand Canyon rapids.

---

140. A bill to authorize the licensing of Hualapai Dam, and for other purposes, HR 3167, 97th Cong. (April 8, 1981), http://140.147.249.9/cgi-bin/bdquery/D?d097:1:./temp/~bdd86O:@@@L&summ2=m&l/bss/d097query.html#rel-bill-detail (accessed March 20, 2006).
141. Langbein, "L'Affaire LaRue" (1983):45.
142. La Rue, *Water Power and Flood Control*, Plate 3.

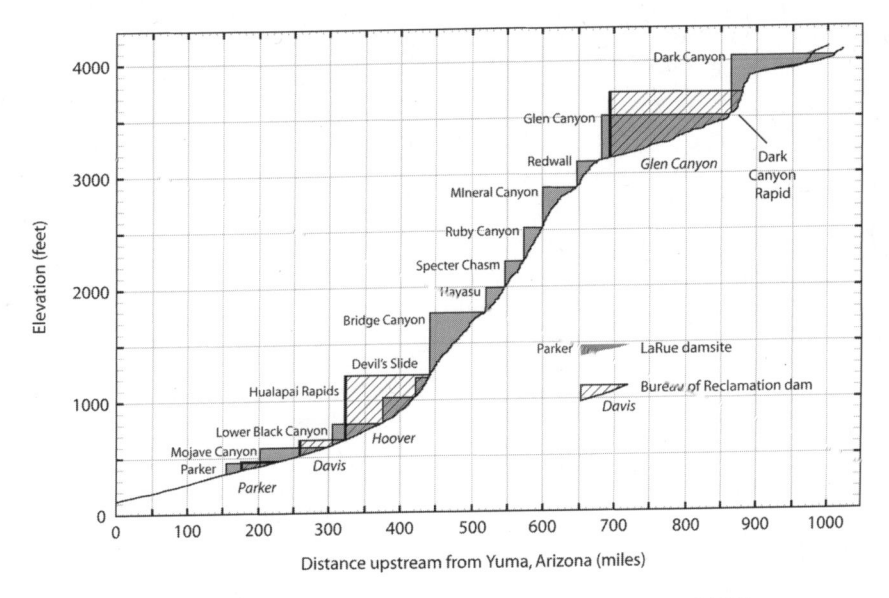

Proposed dam sites along the Colorado River (see page 263 for locations of dam sites in Grand Canyon). Had La Rue's plan been implemented, almost the entire course of the river would have been impounded behind dams with almost no free-flowing water.

A Legacy Remembered

Few people now remember Eugene La Rue and his battles with the Bureau of Reclamation and other water interests that opposed his plans for the Colorado River. Claude Birdseye is remembered on lists of top engineers and on national park monuments. Perhaps a handful of people remember Roland Burchard, Herman Stabler, and Frank Dodge. Among geologists, Raymond Moore has a place of high respect, but not for his contributions to the locations of dam sites on the Colorado River. Lewis Freeman's books are fixtures on the used-book market for Grand Canyon and river memorabilia, Lint is largely forgotten, and Blake's river career is the subject of an obscure biography. Kominsky's smiling visage appears in old publications, but almost nobody recalls Frank Word. Emery Kolb is mostly remembered for his long photographic career and pioneering river running with his brother, not his contributions to the 1923 USGS expedition.

There is a large, ever-growing contingent who may not know who these people were, but they certainly use and benefit from their contributions to our knowledge of the Colorado River in Grand

Canyon. Whitewater recreationists, in numbers ranging from 25,000 to 30,000 persons annually, run the Colorado River from Lee's Ferry at least down to Diamond Creek. Thousands more make the short-distance trip from Diamond Creek down to the various takeouts on Lake Mead. Most of these river runners, and certainly their guides, have in their possession one valuable document: an accurate river guidebook.

In their survey and mapping, Burchard, Birdseye, and the rest of the crew laid the foundation for future Grand Canyon river running with their painstakingly created plan-and-profile maps. By publishing those maps, a Grand Canyon trip could no longer be considered an expedition into the unknown or barely known; there no longer was any guessing as to the location of Separation Rapid. In 1962, canoeist Les Jones produced the first published river guide, a continuous scroll based on the USGS plan-and-profile maps that included the surveyed drops through individual rapids.[143] Seven years later, river runner and graphic artist Buzz Belknap introduced the illustrated, annotated waterproof book-style guide that would become the industry model, also based on the USGS maps.[144] Another popular guidebook uses the river miles and drops through rapids measured by the 1923 expedition.[145]

On the one-hundredth anniversary of the Powell expedition, hydrologist Luna Leopold paid tribute to the 1923 USGS expedition when he chose to analyze their painstakingly surveyed longitudinal profile as part of his discussion of the geomorphology of the

---

143. In 1953, Leslie Allen Jones became the first person to take a canoe down the Grand Canyon; ten years later, he repeated the trip on the extreme low water flows while Lake Powell was filling. Both times he was alone. Webb, Melis, and Valdez, *Observations of Environmental Change*, 3. Jones sold copies of his scroll map, which included his extensive annotations. Leslie A. Jones, untitled Grand Canyon river map, 1962.

144. Buzz Belknap, *Grand Canyon River Guide* (Salt Lake City, UT: Canyonlands Press, 1969). Buzz Belknap was a veteran river traveler, having participated in both the only complete uprun of the Colorado River through Grand Canyon (in jet boats in 1960) and a run in tiny plastic "sportyaks" in 1963, taking advantage of the same low flows as Les Jones. Buzz worked with his father, the accomplished photographer William Belknap, and sister, Loie Belknap Evans, to produce the guide, which has been revised and updated many times since, and is still in print. For more information on the Belknaps, see the 275-276Cline Library online exhibit "Bill Belknap, Photographer," http://www.nau.edu/library/speccoll/exhibits/belknap/ (accessed March 21, 2006). For a thorough discussion of the history of Grand Canyon river maps and guides, see Richard Quartaroli, "Evolution of the Printed Colorado River Guide in Grand Canyon, Arizona," in Cline Library *A Gathering of Grand Canyon Historians:* Anderson, ed., 155–62.

145. Stevens, *The Colorado River in Grand Canyon*. Until 1996, Stevens also included the USGS rapid drop measurements.

Colorado River in Grand Canyon.[146] He calculated a statistic that was widely used to describe the river's energy: 50 percent of the drop occurred in 9 percent of the distance through Grand Canyon. He also noted some peculiar deviations of that profile from a straight line, spurring a twenty-first century analysis of profile convexities, large and small, that characterize a river strongly affected by deposition from side canyons.[147] One of the startling conclusions is that substantial alluvial fill underlies the Colorado River, indicating that Grand Canyon currently is filling up, albeit slightly and slowly, in the present climatic regime of the Holocene.

At the beginning of the twenty-first century, development of satellite-based platforms and computer-driven image analysis, as well as light detection and ranging (LIDAR), would replace the manual surveying techniques that Birdseye pioneered. The profile of the Colorado River, so diligently surveyed in 1923, would be replicated using airborne LIDAR in 2000.[148] One conclusion follows Leopold: in 2000, 66 percent of the drop through Grand Canyon occurred in 9 percent of the distance, indicating that the rapids may becoming more severe in the face of lowered releases by Glen Canyon Dam compared with the unregulated river. What once was used to promote dam construction is now used for environmental monitoring. Perhaps the least recognized of Birdseye's achievements, as well as those of his surveyors (especially Burchard) is the instrumental survey of the Colorado River, which ultimately yielded a longitudinal profile from its headwaters in Colorado to the Sea of Cortés.[149]

In the April 1927 issue of the *Geographical Review*, another aspect of La Rue's career was published. He coauthored an article with USGS geologist Kirk Bryan entitled "Persistence of Features in an Arid Landscape: The Navajo Twins, Utah."[150] The article contains

---

146. Luna B. Leopold, "The rapids and the pools—Grand Canyon" (Washington, DC: The Colorado River Region and John Wesley Powell, U.S. Geological Survey Professional Paper 669, 1969), 131–45.
147. Thomas C. Hanks and Robert H. Webb, "Effects of tributary debris on the longitudinal profile of the Colorado River in Grand Canyon," *Journal of Geophysical Research*, vol. III, F02020, doi:10.1029/2004JF000257, 2006.
148. Magirl, Webb, and Griffiths, "Changes in the water surface profile."
149. Perhaps the best known of these surveys is U.S. Geological Survey, *Plan And Profile of Colorado River From Lees Ferry, Ariz., to Black Canyon, Ariz.–Nev. and Virgin River, Nev.* (Washington, DC: U.S. Geological Survey, map folio, 1924). However, published map folios for everything from the Green and San Juan rivers to the Salt River in Arizona also exist.
150. Kirk Bryan and E. C. La Rue, "Persistence of Features in an Arid Landscape: The Navajo Twins, Utah," *Geographical Review* 17 (April 1927): 251–57.

repeat photography of the prominent landmark on the north side of Bluff, Utah, with originals from 1875 and 1909, and 1925 matches by La Rue. La Rue was one of the first scientists to use repeat photography as a scientific tool in the United States, following Kolb, who was the first to purposefully match another's photograph in the region. We—the authors of the book you are holding in your hands—have specialized in this particular technique and have matched hundreds of photographs originally taken by La Rue.[151] He used photography to document reservoir capacities; we matched his images to document environmental change; ironically much of which was caused by the presence and operation of Glen Canyon Dam, built about eleven miles upstream from La Rue's proposed high dam.

La Rue's water-development plan would have been an environmental nightmare for the Colorado River, but environmental concerns were not a factor in any of the deliberations of the 1920s. As evident in both the Echo Park controversy and the attempts to dam the river in Grand Canyon, environmental concerns ultimately became paramount, especially in the last third of the twentieth century. In response to the large diurnal fluctuations in flow releases from Glen Canyon Dam, and convinced in part by repeat photographs showing environmental damage (particularly beach erosion), in 1992, Congress, led by Senators Dennis DeConcini and John McCain from Arizona, passed the Grand Canyon Protection Act.[152] Ostensibly, this law clearly states the need to protect the riverine resources of Grand Canyon from dam-related degradation. The era of dam sites and dam construction in Grand Canyon ended, but the legacy of the 1923 USGS expedition will live on as long as river runners ply the waters of the Colorado River.

---

151. Publications by USGS scientists using repeat photography of La Rue images include Turner and Karpiscak, *Recent Vegetation Changes*; Melis and others, *Magnitude and Frequency Data*; Bowers, Webb, and Rondeau, "Longevity, recruitment, and mortality"

152. Grand Canyon Protection Act of 1992, bill H.R. 429, 102nd Congress, http://rs9.loc.gov/cgi-bin/query/F?c102:23:./temp/~mdbsG7SScF:e239896; (accessed March 20, 2006).

About the Authors

Diane Boyer is an archivist working for the U.S. Geological Survey overseeing the Desert Laboratory Collection of Repeat Photography in Tucson, Arizona. She has a degree in animal health science (B.S., University of Arizona, 1983). In 1986, she began working as a photo archivist at the Arizona Historical Society; later, she joined the staff of Northern Arizona University's Cline Library Special Collections and Archives Department. She grew up hearing her grandfather's stories of life as a USGS stream gager in the Grand Canyon in the 1920s. In 1991, she made her first Grand Canyon river trip as a field assistant to Robert Webb, and has returned to the canyon many times since. She has published several articles, most of which deal with Grand Canyon and Colorado River history. This is her first book.

Robert Webb has worked on long-term changes in natural ecosystems of the southwestern United States since 1976. He has degrees in engineering (B.S., University of Redlands, 1978), environmental earth sciences (M.S., Stanford University, 1980), and geosciences (Ph.D, University of Arizona, 1985). Since 1985, he has been a research hydrologist with the U.S. Geological Survey in Tucson and an adjunct faculty member of the Departments of Geosciences and Hydrology and Water Resources at the University of Arizona. Webb has authored, co-authored, or edited nine books, including *Environmental Effects of Off-Road Vehicles* (with Howard Wilshire); *Grand Canyon, A Century of Change; Floods, Droughts, and Changing Climates* (with Michael Collier); *The Changing Mile Revisited* (with Raymond Turner); *Cataract Canyon: A Human and Environmental History of the Rivers in Canyonlands* (with Jayne Belnap and John Weisheit); and most recently, *The Ribbon of Green: Long-Term Change in Woody Riparian Vegetation in the Southwestern United States* (with Stanley Leake and Raymond Turner).

Index

A

Abejos (Bee) River, 29
aerial photography, 246
Aerotopograph Corporation of America, 246
aircraft, 193, 193n39
Alamo River, 25, 27
All-American Canal, 41, 44
American Geographical Society, 245
American Society of Civil Engineers, 247
American Society of Photogrammetry, 247, 254
Andrade, Guillermo, 24, 26
Andrus Wash, 197–98
annual flow volume, present-day, 46
archaeological sites, 117–18, 172, 172n10, 173, 193, 199
Arizona, 71, 266, 273; and Colorado River Compact, 45–47
Arizona Engineering Commission, 54, 71
Army Reserve Engineer Corps, 247
Ashurst, Henry Fountain, 138n18
Ashurst, William H., 138, 138n18
Associated Press, 144

B

Badger Creek Rapid, 98–101, 143
Barry, Fred T., 107n26
Bass, Edith, 149n37
Bass, William Wallace, 160n51
Bass Cable Crossing, 160, 161, 161n54, 163, 163n55
Bass Canyon Rapid, 161
Bass Trail, 142, 158, 160
Beach, Rex, 161, 161n53
Bedrock Rapid, 169–71
Belknap, Buzz, 276, 276n144
Bert's Canyon, 113
birds: blue heron, 197–98; crane, 127; duck, 176, 198, 231, 235–36; quail, 198, 232, 235
Birdseye, Claude, 8, 54–56, 80, 83–85, 87–89, 206, 211, 244, 255–56; expedition journal, 3–4, 98, 100, 102–4, 107, 109–13, 117–19, 127, 130–31, 133–36, 138–39, 142, 146, 151–55, 157–63, 166, 169–76, 178–81, 183–88, 190–91, 194–95, 198–204, 213, 221, 223–24, 227, 229; post-expedition years, 245–47, 259–60; profile, 58–60; resignation from USGS, 245–46; return to USGS, 246
Birdseye, Roger, 60, 86, 129, 141, 151, 160, 180, 204–5, 239, 258
Black Canyon, 55, 57, 57n34, 75, 235, 237, 255, 265–66
Blacktail Canyon, 166, 167n3
Blake, Elwyn, 52–54, 85, 255–57; autobiography, 101, 131, 141, 180, 227–28, 236n19, 238; expedition journal, 3–5, 98, 102–5, 110, 112–13, 116–17, 120, 128–29, 134, 138, 142–43, 149–51, 156–59, 162–63, 166–70, 172–74, 177, 179, 181–86, 188, 190–93, 197–98, 203–5, 211, 213–15, 217–18, 224–26, 233–34, 236–40; letters, 126; post-expedition years, 252–53; profile, 67–68
Blake, William Phipps, 20–22, 24
Blythe, Thomas Henry, 24
boating technique: Major Powell move, 107; nosing, 103n18
boats, 79–80, 91–92, 94, 94n3, 106n22, 108–9, 244; damage and repairs, 133–34, 142, 150, 181–82, 184, 191, 212, 226
Boucher Rapid, 155
Boulder Canyon, 44–45, 48, 57, 57n34, 75, 235, 248, 265–67; 1922 expedition, 54–56
Boulder Dam, 30n57, 41n29, 75, 82, 246, 246n18, 266, 268, 270, 272. *See also* Hoover Dam and La Rue high dam
Boulder Narrows, 105
Boulder Rapid (President Harding Rapid), 113, 113n33
Bridge Canyon, 216, 265, 271–73
Bridge Canyon Rapid, 215–16

280

Index

Bridge of Sighs, 113
Bright Angel Creek, 138-40
Bright Angel Hotel (Lodge), 141n25
Bright Angel Rapid, 261
Bright Angel Restaurant, 141
Bright Angel Trail, 140, 142, 148
Brower, David, 273–74
Brown, Frank, 49–50, 103n19
Brown-Stanton expedition, 109n29, 114
Bruce, Eddy (guide), 133
Bryan, Kirk, 277
Buckfarm Canyon, 113
Burchard, Roland, 80, 97, 211, 229, 232, 244; handwritten notes, 213; letters, 3, 6, 142–43; post-expedition years, 247–48; profile, 73–75
burros, feral, 192n38, 203
burro trails, 192n38

C

California, 34, 265; and Colorado River Compact, 45–47; Irrigation District Act (1887), 29. *See also* Los Angeles
California Development Company (CDC), 25–28, 32. *See also* Imperial Irrigation District (IID)
Calloway, O. P., 24
cameras, 80–81, 81n107, 81n108, 81n109, 102n15. *See also* motion pictures
Cameron, Ralph, 140n23, 140n24, 259
Canyon Copper Company, 128n7
Cape Solitude, 118–19
Carico, Nellie, 167n2
Carnegie Institute-California Institution of Technology Expedition, 250
Cataract Canyon, 252; 1921 expedition, 52–53
Cataract Creek, 183
Cave Springs Rapid, 108–9
Central Arizona Project, 272
Chaffey, George, 25–28
Chandler, Harry, 26, 30, 265
Chenoweth, William, 52–53

Church of Jesus Christ of Latter-day Saints. *See* Mormon settlers
Clear Creek, 136, 138
Cogswell, Raymond C., 241n24
Colorado, 34; and Colorado River Compact, 45–47
Colorado River: changes in rapids, 277; early explorers, 48–50 (*see also* Grand Canyon expedition of 1923); floods, 19 (*see also separate entry*); as free-flowing river, 18–23; as fully regulated river, 273; uses of, 46–47; water development, 13, 23; water use, 11, 14–16. *See also specific place names*
Colorado River Aqueduct, 272
Colorado River Compact, 45–47, 55
Colorado River Storage Project, 56, 264–65, 273
Colter, Mary Jane, 123n3, 139n22
Compact Point, 11, 45, 264
Connecticut River, 34
Conquistador Aisle, 166, 168
Cory, H. T., 27–29
Cottonwood Canyon, 136–38
Cove Canyon, 184
Cowan, John F., 245n12
Crash Canyon, 120, 120n46, 121, 123
Cremation Creek, 138
Crosby, Walter W., 151, 151n40
Crystal Creek, 155
Crystal Rapid, 155n47, 261

D

Daggett, John S., 82, 82n116
dam sites, 42, 136, 138, 159, 181; identification of, 13–14, 48, 56, 93, 196, 206; and La Rue's proposal, 263–68. *See also names of dams, and place names*
Dark Canyon, 53
Dark Canyon Rapid, 274
Davis, Arthur Powell, 3, 36–41, 43–44, 48, 54–57, 66, 82, 255, 263, 267, 270, 272

Davis, D. W., 55–56
Davis Dam, 272–73
Deer Creek, 174–75
Deer Creek Falls, 173n13
Dellenbaugh, Frederick, 50
Desert View, 123, 123n3
Devil's Slide and Devil's Slide Rapid, 230–32
Diamond Canyon, 211
Diamond Creek, 46–47, 55, 142, 200–202, 204–9, 211
Diamond Creek Rapid, 211–13
Diamond Peak, 203
diaries of 1923 expedition, 3–8. *See also names of expedition members*
Dickinson, William E., 269–70
Dimock, Brad, 215n3
Dodge, Frank, 94–97, 107, 121, 178–79, 217–18, 218n5, 249–52; autobiography, 86–87, 250; expedition account, 3, 108, 178–79; letters, 153–54, 271; profile, 75–78
Doris Rapid, 175, 175n14
driftwood: burning, 116, 116n39, 124–25, 185; measureing, 215
Drift-Wood Burners (DWBs), 116n39
drop of rapids, measuring, 102, 102n17, 175n14
drought: early twentieth century, 26; present-day, 12–13, 274
Dubendorff, Sylvester, 171n8
Dubendorff Rapid, 171–72, 171n7, 171n9
Dudley, Donald, 139–40

E

Echo Park, 273
Eddy, Clyde, 102n16, 167n2
85 Mile Rapid, 138
Elliott, Herman R., 167, 167n2
El Tovar Hotel, 141, 141n29
Elves Chasm, 166–67
Embudo, New Mexico, 34
environmental movement, 273–74
European contact, 15
Evans, Richard T., 158
expedition of 1923. *See* Grand Canyon expedition of 1923

F

Fairchild Aerial Surveys, 250
Fall, Albert, 43

Fall Canyon, 200
Fall-Davis Report, 44, 57, 265, 267
Fang Rocks, 214n3
Farlee Hotel, 207n49
Federal Power Commission, 44, 46
Fellows and Stewart Shipbuilding Works, 79, 109
Fence Fault, 109
Fern Glen Canyon and Rapid, 183–84
fire: driftwood, 116, 116n39, 124–25, 185; forest, 231
fish: bonytail (humpback chub?), 184, 184n29
Fishtail Canyon and Fishtail Canyon Rapid, 175
Fisk, Charles, 82, 82n113, 180, 204–5
Flagstaff, Arizona, 85–87
Flavell, George, 49, 56, 171n9, 185n32, 224n12
flood control, 31, 39–40, 44
floods and flooding, 19, 30–31; of 1905, 27–28; of 1923, 182n27, 185, 187–95, 204, 208, 235
flow regulation, 31, 39–40; advocacy, 32–33
food, 81–83
Forster Rapid, 168
Fort Colville, 236
fossil finds, 113
Fossil Rapid, 168
Fox Movietone News Company, 85, 146–47, 154–55, 239
Fred Harvey Company, 63, 78, 142
Freeman, Lewis, 54–56, 79, 82, 85, 209, 240–41, 244, 253; article, 146–48; book *Down the Grand Canyon*, 188, 255–56; diary, 219–20; expedition journal, 3, 5, 8, 98–107, 109, 112–14, 118–20, 123–25, 127–33, 135, 139–42, 144, 152, 155, 158–61, 163, 167, 169, 174–75, 179, 181–82, 185–88, 190–92, 194, 196, 198, 200, 202, 205, 207–8, 211, 213, 216–21, 229–35, 237, 239; post-expedition years, 254–57; profile, 64–66, 86–87
Furnace Flats, 121

G

gaging stations, 34
Galloway, Nathaniel, 49, 79, 155n46
Galloway, Parley, 102n16
Galloway Canyon, 171n7

Gannett, Henry, 36
García López de Cárdenas, 56
Garfield, James R., 38–39
Gass, Edna, 196
Gateway Rapid, 184
gear, 80
geological formations: Bright Angel Shale, 113; Chinle Formation, 273; Muav Limestone, 113; Navajo Sandstone, 273; Tapeats Sandstone, 167, 186. See also Pa Snuff
geology of Colorado River, 18–23
Gila River, 46; floods, 19, 30; water development, 23
Gilliland, Dick, 129, 129n8
Gilpin, William, 16
Girand dam site, 46, 55, 196, 205–6, 216, 273
Girand, James, 46, 55
Glen Canyon, 42–43, 248, 264; 1922 expedition, 54–56
Glen Canyon Dam, 11–12, 273
Gneiss Canyon and Gneiss Canyon Rapid, 216, 273
Goldwater, Barry, 117n39, 261
Grand Canyon, 119; environmental concerns, 277–78; Western Canyon expedition (1922), 54–56
Grand Canyon Association, 262
Grand Canyon expedition of 1923: boatmen (see Blake, Elwyn; Freeman, Lewis; Lint, Leigh); boats, 79–80, 91–92, 94, 94n3, 106n22, 108–9, 133–34, 142, 150, 181–82, 184, 191, 212, 226, 244; cook (see Kominsky, Felix; Word, Frank); endpoint, 238–41; equipment, 79–83, 166; geologist (see Moore, Raymond); head boatman (see Kolb, Emery); hydraulic engineer (see La Rue, Eugene C.); illness and injury, 123–24, 126–29, 150, 183, 200, 206–8, 212–13; interpersonal conflict, 84, 87, 120–22, 126, 152–54, 164, 209, 240–41, 244, 255–57; launch, 97–98; monuments to, 245; preparations, 56–58, 85–89; publicist (see Freeman, Lewis); resupply, 81–83, 85, 129–30, 133, 146, 160–61, 180, 200, 202; river accidents, 93, 107, 217–21; rodman (see Dodge, Frank); survey work, 229–30, 230n16, 275–78 (see also place names); expedition members, 58–78;
topographer (see Burchard, Roland); topographic assistant (see Stabler, Herman); trip leader (see Birdseye, Claude); weather, 100, 116, 121, 124, 130, 156–57, 160, 164, 181. See also names of expedition members, and subject entries
Grand Canyon longitudinal profile, 118n41, 276–77, 277n147
Grand Canyon National Park, 57, 63, 117, 138n18, 141–46, 173, 259–62
Grand Canyon National Park Foundation, 244
Grand View Point, 128–30
Grand Wash, 142, 233–34
Grand Wash Cliffs, 233–34
Granite Gorge, 133–36, 138n20, 143, 212n1
Granite Narrows, 173n13, 174
Granite Park, 199
Granite Park Canyon, 200
Granite Rapid, 151
Granite Spring Rapid and Granite Spring Canyon, 203
Grapevine Camp, 135
Grapevine Creek and Rapid, 136
Green River, 52, 252, 273; 1922 expedition, 53–54
Green River, Utah, 34
Gregory, Herbert E., 73
Grover, Nathan, 69–70, 88, 263, 265

H

Hamblin, Kenneth, 193n40
Hance, "Captain" John, 123n2
Hance Rapid, 123n2, 127–33, 131n11, 132n13, 143, 253
Hance Trail, 118, 127–28
Hanlon Heading, 25–26
Hansbrough, Peter, 50, 114, 114n37
Harding, President Warren G., 101
Harriman, Edward H., 27
Havasu Creek, 142, 179–80, 182–83
Havasupai Indian Reservation, 179
Hayes, President Rutherford B., 33
Headgate Rock Dam, 273
Hearst, William Randolph, 128n7
Hell Diver Rapid, 232
Hermit Creek and Hermit Creek Rapid, 151, 154–55, 239
Hermit Trail, 142
Hohokam water use, 14
Holmstrom, Buzz, 250

Hoover, Herbert, 45, 267
Hoover Dam, 3, 272. *See also* Boulder Dam and La Rue high dam
Horn Creek Rapid, 150–51, 159n50
House Rock Rapid, 104–5, 104n20
House Rock Wash, 104
Hualapai Indians, 208
Hunt, George W. P., 141, 141n28
Hurricane fault zone, 196–98
Hyde, Glen and Bessie, 214n3, 261
hydroelectric power, 44–45, 47, 264–66, 272–73. *See also* Southern California Edison and Utah Power and Light Company

I

Iceberg Canyon, 234
illness and injury, 123–24, 126–29, 150, 183, 200, 206–8, 212–13
Imperial County, California, 29–31
Imperial Dam, 273
Imperial Irrigation District (IID), 29–32, 41
Imperial Valley, 24–28, 41
Indian Canyon, 199
Indian Gardens, 140, 140n23, 145
insects and spiders, 169, 169n4; bees and butterflies, 176; flies, 169, 169n4; horsefly, 202; mosquitoes, 202, 237; scorpions, 196; tarantula, 207
international boundary, U.S.-Mexico, 40–41
interpersonal conflict, 84, 87, 120–22, 126, 152–54, 164, 209, 240–41, 244, 255–57
irrigation, Native Americans and, 14–15
Irrigation Congresses, 37

J

Jackson, William Henry, 51
Jones, Leslie Allen, 276, 276n143
Jumbo Wash, 235–36
Junction Dam, 51

K

Kaibab Plateau, 114n38
Kanab Creek and Kanab Creek Rapid, 176–77
Kansas State Geological Survey, 73, 251–52
Kendrick, Frank C., 49

Kettner, William, 41
Keyhole Canyon (140 Mile Canyon), 175. *See also* Neighing Horse Canyon
KFI radio station, 231
KHJ radio station, 82–83, 118, 147, 172–73, 215, 231, 243
King, Clarence R., 33
Kittner, James, 149, 149n38
Kolb, Blanche Bender, 64
Kolb, Edith, 64, 84, 94, 129–31, 131n11, 140, 151, 153, 181, 253; diary, 67–68, 78
Kolb, Ellsworth, 50, 52, 60, 62–63, 85, 130, 133, 136, 140, 149n37, 223
Kolb, Emery, 50, 52–53, 79–80, 85, 136, 140, 149n37, 155n45, 209, 217, 217n4, 223; expedition journal, 3, 6–7, 98, 100–102, 106, 110, 116, 118, 120, 126–27, 129–31, 133–34, 136–38, 161; *Grand Canyon Film Show,* 258–59; interpersonal relations, 84–85, 87, 152–54, 228, 228n15, 244, 248, 252, 255–57; letters, 150, 180–81, 205–8, 240–41; as medic, 127, 183; post-expedition years, 258–62; profile, 60–64
Kolb Brothers Photo Studio, 62–63, 140–41, 141n25, 152, 259, 261–62
Kominsky, Felix, 180, 180n22, 209, 238–39; post-expedition years, 258; profile, 78
Kwagunt Rapid, 118–20, 119n43

L

Laguna Dam, 28, 273
Lake Cahuilla, 14, 21–22, 25–26, 28
Lake Havasu, 272
Lake Mead, 209, 223n11, 250, 274
Lake Mohave, 272–73
Lake Powell, 12–13, 274
Langbein, Walter, 270–71, 274
La Rue high dam (Lee's Ferry), 42–43, 71, 248, 263–65, 273, 278. *See also* Boulder Dam, Hoover Dam
La Rue, Eugene C., 3, 40, 78, 81–82, 85, 93, 152–54, 244, 249–50; Colorado River dam proposal, 263–68; Colorado River monograph, 263–65; Colorado River research, 42–43, 46–47, 71; expedition journal, 3, 5–6, 100–102, 105–7, 114–18, 123–24, 126–30, 134, 138,

180, 221; and 1921 Cataract Canyon expedition, 52–53, 53n24; and 1922 expedition, 54–56; post-expedition years, 263–72; profile, 69–72; resignation from USGS, 269–71; use of repeat photography, 278; water-development plan, 8
La Rue, Mabel, 71, 71n73, 81n106, 144n32, 148n36, 195; typescript, 144–45, 148–49
Las Vegas, Nevada, 44
Lauzon, Hubert R. "Bert," 149, 149n37, 150
Lava Canyon Rapid, 123
Lava Cliff Rapid, 209, 223–28, 223n11, 224n12, 250
Lava Falls, 190n35
Lava Falls Rapid, flood at, 185–94, 185n31–185n32
Lee's Ferry, 43, 54, 87–88, 90–92
legislation, federal: Boulder Canyon Project Act, 47; Carey Act, 37; Federal Water Power Act, 44, 69; Grand Canyon Protection Act, 278; Kinkaid Act, 41; Land Ordinance of 1785, 17; National Historic Preservation Act, 262; Reclamation Act (Newlands Act), 37–38
Leighton, M. O., 71
Leopold, Luna, 276–77
LIDAR (light detection and ranging), 277
Lint, Leigh, 52–54, 93, 209, 255–57, 261; expedition journal, 3, 6, 104, 117, 124, 126, 129–31, 136, 142, 148, 152, 155, 157, 159–60, 163, 171–73, 176–78, 182, 184, 186–87, 191, 194, 196, 199–202, 206, 208, 211–12, 214, 216–17, 221–22, 224–25, 227–29, 231–32, 235–38, 240; post-expedition years, 253–54; profile, 66–67
Little Colorado River, 118–20, 120n44, 190
Little Colorado River Confluence, 122–23, 261
Little Nankoweap Creek, 117
Litton, Martin, 273
livestock, 229
logistics, 79–83, 86, 89
Lonely Dell Ranch, 92–93
Loper, Bert, 49–50, 52–54, 53n24, 60–62, 61n44, 67, 103n19, 161n54, 252
Los Angeles, California, 26, 44–45, 265, 272
Lost Creek Rapid, 229
Lower Basin (Colorado River), defined, 45–46

M

malaria, 202n44
Manifest Destiny, 15–16, 15n11
Marble Canyon, 87–88, 97–98, 101, 106–7, 119, 143, 273
Marston, Otis "Dock," 4, 153, 214n3, 252–53, 261
Matthes, Francois E., 158
Mendenhall, Walter, 195
Meriwhitica Canyon, 223n11
Mexicali Valley, 26, 30
Mexican Revolution, 40
Mexico, 26, 30, 46
Mineral Canyon and Creek, 129–30
miners and mining, 123n4, 128n7, 160n51, 185n30
mines: Grandview copper mine, 128n7; Katherine Mine, 195; Last Chance copper mine, 128n7; Little Chicken mine, 185, 185n30
Miser, Hugh, 52, 73
Moeur, Benjamin B., 47
Mohave Canyon, 264
Mohawk Canyon, 184
Mojave Indians, 15
Montéz, Ramon, 49
Monument Creek, 151
Moore, Charles, 30–31
Moore, Raymond, 85, 244; expedition journal, 3, 7; interview, 221; letters, 166, 170; post-expedition years, 251–52; profile, 72–73
Morelos Dam, 273
Mormon settlers, 15, 17, 196
motion pictures: Fox News and, 146–47, 151, 153, 239; Kolb and, 93, 100, 100n12, 106, 116, 122, 131n11, 153, 155n45, 161, 179; Kolb brothers and, 62–63, 85, 142; La Rue and, 93, 112, 116, 122, 138, 151, 153, 218n5
Mulholland, William, 44, 265–68

N

Nankoweap Creek, 117–18
Napoleon's Tomb, 235
National Canyon, 183
National Geographic Society, 259
National Irrigation Association, 37
National Park Service, 63, 140n23, 161n54, 163n55, 192n38, 261–62. *See also* Grand Canyon National Park
Native Americans and water use, 14–15
Navajo Bridge, 93, 100n11
Needles, California, 238–39
Neighing Horse Canyon, 175. *See also* Keyhole Canyon
Nevada, 45–47
Nevills, Doris, 175n14
Nevills, Norman, 52, 52n19, 117n39, 175n14, 261
Nevills Rapid, 126–27
Newell, Frederick Haynes, 34–39
Newlands, Francis, 37. *See also* legislation, federal
New Mexico, 45–47
New River, 27
newspapers and periodicals, 242–43; *Arizona Gazette*, 144; *Deseret News*, 83; *Engineering News-Record*, 243; *Geographical Review*, 277; *Literary Digest*, 83, 135; *Los Angeles Times*, 26, 32, 82–83, 195, 205, 242–43, 265–66, 269; *Mining Congress Journal*, 243; *National Geographic*, 254–55; *San Juan Record* (Monticello, Utah), 4, 4n11; *Scientific American*, 247; *Sunset Magazine*, 240, 255; *Washington Post*, 242
Nixon Rock, 157
Noble, Levi, 73
North Canyon Rapid, 105–6
Norviel, W. S., 55

O

Ockerson, John, 29
Oliver, Norman, 167, 167n2
128 Mile Rapid, 168
164 Mile Rapid, 182
194 Mile Canyon, 197
O'Sullivan, John L., 15n11
Otis, Harrison Gray, 26
Owens River, 26, 44

P

Pahl, Catharine, 86, 93–94, 98
Paige, Sidney, 52–53, 72–73
Palo Verde Dam, 273
Parashant Wash, 198
Paria Riffle, 98–99
Parker Dam, 272
Pa Snuff (Bright Angel Shale), 201, 203. *See also* geological formations
Pathé-Bray movie trip, 249–50, 252, 261, 269, 271
Peach Springs Wash, 207
Pearce, Harrison, 231n17
Pearce Ferry, 231, 231n17, 233
Phantom Camp, 145
Phantom Ranch, 139, 139n22
photogrammetry, 246
photography, 62–63, 81, 84–85. *See also* Kolb Brothers Photo Studio; motion pictures
Pipe Creek and Pipe Creek Rapid, 139, 146, 149, 151
Point Hansbrough, 114
Powell, John Wesley, 14, 17n18, 48–50, 112n32, 217, 223, 235, 270; as director of USGS, 2–3, 17, 33–35; irrigation plan, 16–18
President Harding Rapid, 113–14, 113n33, 114n35–114n36, 117
prior appropriation doctrine, 45
Prospect Canyon, 185–86
Prospect Canyon No. 2, 196
Prospect Creek, 194
Prospect Creek No. 3, 197
prospectors, 59–60, 138n18, 203
public domain, states and, 37
publicity for Grand Canyon expedition, 57–58
Pumpkin Springs, 201

R

radio, 82–83. *See also* KHJ radio station
Randolph, Epes, 27
rapid, defining, 104. *See also names of rapids*
rapids, running *vs.* portage, 83
Red Canyon, 127, 134
Redwall Cavern, 112, 112n32
Reeside, John, 53–54
remote sensing, 246
renumbering, of rapids and river miles, 122–23

repeat photography, 278
reservoir sizing, 35
resupply, 81–83, 85, 129–30, 133, 146, 160–61, 180, 200, 202. *See also* Birdseye, Roger; Fisk, Charles
Richards, Henry, 50
Richmond, William, 49
Rio Grande, 34
river accidents, 93, 107, 217–21
river expeditions, 48–50. *See also* Grand Canyon expedition of 1923
river guides, 81, 81n110, 276
river rise and fall, 126, 136, 147, 156–57, 159–60, 162, 164, 167, 181–82, 185, 187, 226, 226n13. *See also* floods and flooding
Roaring Twenties rapids, 106–7, 107n25
Rockwood, Charles Robinson, 24–28
Roosevelt, Theodore, 37
Royal Arch Creek, 166–67
Ruby Creek Rapid, 159
Russell, Charles, 161n54
Rust, David, 94, 94n5

S

Saddle Canyon, 117
Salt Creek Rapid, 151
Salton Sea, 11, 25n41, 28
Salton Sink, 20–23, 27
San Andreas Fault, 20
San Fernando Valley, 26
San Juan River, 120n45; expedition of 1921, 51–52
Santa Cruz River, 14
Santa Fe Railroad, 63, 142
Sapphire Canyon and Sapphire Canyon Rapid, 158
Scattergood, E. F., 44
Sea of Cortés, 20
semaphore, 183, 183n28, 225–26
Separation Rapid, 209, 216–21
Serpentine Creek and Serpentine Creek Rapid, 159–60
75 Mile Canyon, 126
Sheer Wall Rapid, 104
Shinumo Creek and Rapid, 162–63
Shurtliff, J. P., 146–47, 146n34, 151
Sierra Club, 273n139
60 Mile Rapid, 118
Smith, G. E. P., 263
Smith, George Otis, 39, 41–42, 268–69
Smithsonian Institution, 244

Snake River, 52
Soap Creek and Soap Creek Rapid, 101–3, 102n16, 143
Sockdolager Rapid, 130, 135, 143
solar eclipse, 175, 175n15
Southern California Edison, 43, 45, 48, 79, 82; Big Creek Powerhouse No. 1, 244; as funding source, 52, 56, 267
Specter Chasm, 169
Specter Rapid, 169
Spencer, Charles, 223n11
Spencer Canyon, 223, 223n11
Spencer Creek, 225
Spring Canyon, 198–99
Spring Cave Rapid, 107
Stabler, Herman, 54–56, 69–70, 81, 88; expedition journal, 3, 7, 146, 151, 155, 158, 160, 162–63, 167, 169, 176–77, 182–84, 186, 191, 193–203, 206–8, 212, 215–16, 226, 230–31, 235–36; post-expedition years, 248–49; profile, 68–69
Stairway Canyon, 184
Staker, Sadie, 98n9
Stanton, Robert, 49–50, 109, 112n31, 114, 135n15, 159, 221, 223–24, 224n12
Stanton's Cave, 112, 112n31
states: and irrigation projects, 37; and water rights, 45–47
Stetson, Clarence, 54–56
Stewart, John T., 167, 167n2
Stone, Julius, 49, 223–24, 241n24
Stone Creek, 171n7
Supai Indians, 180
Surprise Canyon and Rapid, 228–29
survey equipment, 80–81
surveying, 88–89
Suspension Bridge, 142, 145
Swing-Johnson bill, 41, 265. *See also* legislation, federal
Swing, Phil, 30, 41

T

Taft, President William, 40–41
Tanner, Seth, 123n4
Tanner Creek Rapid, 124
Tanner Trail, 123n4
Tapeats Creek and Rapid, 166, 171–72, 172n11, 173–74, 173n13
Temple Bar, 235
Temple Butte, 234–35

Ten Mile Rock, 101
36 Mile Rapid, 113
Thompson, Ellen "Nellie" Powell, 98n9
Three Springs Canyon, 202
Thunder River, 173, 173n13
Tiger Wash, 110
topographic mapping, 35, 51, 253–54
tourism, 62, 94, 94n5, 140n23, 142, 151, 160n51, 193n39, 206–7, 207n49, 276
Trail Canyon Rapid, 202
trappers, 49, 107n26
Travertine Canyon and Falls, 212
Travertine Grotto, 213
Trimble, Kelly, 52–54
Trinity Creek, 151
Triple Alcoves, 117
Tristate, Nevada, 237
Triumphal Arch Rapid, 232
Tuckup Canyon, 182
Tuna Creek and Tuna Creek Rapid, 157–58
Turquoise Canyon, 158
29 Mile Rapid, 110
205 Mile Creek and Rapid, 199
209 Mile Rapid, 200, 200n42
217 Mile Rapid, 201–2
220 Mile Canyon, 202
222 Mile Canyon, 202
224 Mile Rapid, 205
231 Mile Rapid, 214
232 Mile Rapid, 214–15
237 Mile Rapid, 216
241 Mile Rapid, 221

U

United States Department of Agriculture, report on Imperial Valley soils (1902), 25
United States Forest Service, 253–54
United States Geological Survey, 33–38, 47; Conservation Branch, 248–49; Division of Engraving and Printing, 246–47; Division of Water Utilization, 71; Hydrography Division, 34; and hydrology, 34–38; and identification of dam sites, 48, 56–57; Kolb's feud with, 259–60; Land Classification Branch, 69; Photograph Library, 262; and Powell's Irrigation Survey, 34, 41–42; river expeditions of 1921 and 1922, 50–56; Topographic Division, 59, 167n2; and U.S. Reclamation Service, 40; Water Resources Division, 3, 42. *See also* Grand Canyon expedition of 1923
United States Public Health Service, 202n44
United States Reclamation Service (U.S. Bureau of Reclamation), 3, 25–26, 33–44, 47–48, 55–57, 204, 235, 237, 267–69; and Colorado River Storage Project, 273–74
United States Senate Committee on Irrigation and Reclamation, 265–67
United States War Department, 34
Unkar Rapid, 124
Upper Basin (Colorado River), defined, 45–46
Upset Rapid, 177–79, 177n16, 195, 209, 219, 251
Utah, 45–47
Utah Power and Light Company, 48, 53

V

Vasey's Paradise, 110–12
vegetation, 167, 176; agave, 201n43; barrel cactus, 213; black willow, 197; cacti, 174, 187; catclaw acacia *(Acacia greggii)* ("Judas Tree"), 113, 113n34; cottonwood, 173; gramma grass, 192; maidenhair fern, 173; mesquite, 113, 215; poison ivy, 175; saw-edged grass, 187; thistle, 185, 187; tule grass, 185; willow, 237
Virgin River, 235
Vishnu Creek, 135
Vulcan's Anvil, 185

W

Walapai Rapid, 234
Walcott, Charles D., 33, 35–36, 38
Waltenberg Canyon, 163
Waltenberg Rapid, 166–67
Waltenburg, John, 163n56
water development, 23–28. *See also* California Development Company (CDC)
weather, 100, 116, 121, 124, 130, 156–57, 160, 164, 181; sandstorms, 121
White, James, 48, 50
whitewater recreationists, 276
Whitmore Canyon and Rapid, 195–96

Widney, Dr. J. P., 22
Widtsoe, John, 54–56
wildlife: bighorn sheep, 169, 172, 179, 179n20, 185; coyote, 232, 235; deer, 113–16, 114n38, 124, 192; feral burro, 192n38, 203; rattlesnake, 128, 191, 198. *See also* birds; insects and spiders
William Grigg Ferry, 234
Woolley, Elias Benjamin "Hum," 49
Woolley, Ralph, 53–54
Word, Frank, 81, 126, 154, 180–81, 195; expedition letter, 7; letter, 181; newspaper interview, 195; post-expedition years, 257–58; profile, 78

Work, Hubert, 55, 206, 268
World War I, 58–59, 65, 74
Worster, Donald, 38
Wozencraft, Oliver Meredith, 24
Wyoming, 45–47

Y

Yuma, Arizona, 34

Z

Zoroaster Rapid, 138

Archean